KU-654-892

Five New World Primates

MONOGRAPHS IN BEHAVIOR AND ECOLOGY

Edited by John R. Krebs and Tim Clutton-Brock

Five New World Primates

A Study in Comparative Ecology

JOHN TERBORGH

Princeton University Press
Princeton, New Jersey

Published by Princeton University Press,
41 William Street, Princeton, New Jersey 08540

In the United Kingdom: Princeton University Press,
Guildford, Surrey

Library of Congress Cataloging in Publication Data will
be found on the last printed page of this book

ISBN 0-691-08337-1
ISBN 0-691-08338-X (pbk)

This book has been composed in Linotron Times Roman

Clothbound editions of Princeton University Press books
are printed on acid-free paper, and binding materials are
chosen for strength and durability. Paperbacks, although
satisfactory for personal collections, are not usually
suitable for library rebinding.

Printed in the United States of America by Princeton
University Press, Princeton, New Jersey

Designed by Laury A. Egan

To the Monkey Watchers of Manu

Contents

Preface

This study would never have been undertaken were it not for our "discovery" of Cocha Cashu in 1973. At that time I had been traveling in Peru and other South American countries for ten years, but had never before seen a place in the lowlands that was utterly pristine. While this may seem surprising in view of the vast expanse of primary lowland forest that remains in Amazonia to this day, the fact is that nearly all of it has been exploited in one way or another, for game, furs, natural rubber, Brazil nuts, prime timber, etc. Animal populations are especially vulnerable to the slightest encroachment, and appear to melt away before seemingly insignificant human populations. This is because of the devastating effectiveness of modern firearms. A skilled hunter equipped with a standard one-shot 16-gauge shotgun can single-handedly eliminate large birds and mammals within a radius of several hours' walk from his dwelling. So, even if the forest looks lush and intact from the air, the appearance is deceptive. Much of it is a hollow shell so far as animals are concerned.

Having seen only places that had been exposed to hunting, I was not fully aware of this. It was the contrast of visiting Cocha Cashu that put all my previous experience in perspective. Animals are actually plentiful, especially primates. Never before had I seen so many monkeys of so many species, much less monkeys that did not flee at the first hint of a human being.

My research to that time had concentrated on birds. The thought of studying primates had never seriously occurred to me, because I had only had a few fleeting glimpses of them rushing off in panic. Interesting creatures perhaps, but impossible to study, I thought. During the first visit to Cocha Cashu I realized that they were not impossible to study after all. And thus was the project born.

Cocha Cashu is not an easy place in which to operate a major research program. It is the way it is by virtue of its extreme remoteness from markets and channels of transportation. One must pay to enjoy the pleasures and benefits of the pristine environment—in time, inconvenience and added expense.

Supplies must be purchased in Cusco, nearly a week away if the trip goes smoothly. Food, fuel, equipment and personnel are loaded onto a chartered truck. We then embark on a two-day journey, crossing two ranges of the Andes before descending more than 3,000 meters onto a verdant plain where several headwater streams gather to form the Rio Alto Madre de Dios. The road is so narrow and winding that traffic is permitted to pass in only one direction—down on Mondays, Wednesdays, and Fridays and up on Tuesdays,

Thursdays, and Saturdays. Even in the best of times the road provides a rather tenuous connection, but during the rainy season it offers no guarantee whatsoever. In 1979 it was severed by landslides and washouts for five months continuously. When it is fully operational, traffic can still be stopped unpredictably by high water in any of the many streams that must be forded.

The road ends unceremoniously at Shintuya, a tiny missionary settlement on the banks of the Alto Madre de Dios. At Shintuya the river substitutes for the road as the only available channel of transport. Everything must be transferred to dugout canoes. These are not ordinary canoes, but work craft 12 to 14 meters long, capable of carrying two tons or more. Ideally suited for river travel, they are powered by outboards or makeshift air-cooled inboards known locally as "pequepeques."

The Alto Madre is a formidable river, swift, broad, and rippled by the play of currents over rocks. Only the most skilled and experienced boatmen run it. One has to know which of many channels to choose, how to read the depth of the water, and how to enter the dozens of rapids. Each time we arrive in Shintuya our friends recount the misfortunes of those who had overturned since our last visit. So far we have been fortunate.

Nearly 100 km downstream, the confluence with the Rio Manu is marked by a line in the water, a sharp boundary between its turbid charge and the sparkling blue of the Alto Madre. Here we turn west and enter a different scene. High muddy banks on one side offset broad sandy beaches on the other. Languid and opaque, the Manu winds through a featureless plain covered by towering forests. Although seemingly innocent in its torpor, the Manu also contains its share of hazards, and is perhaps more insidious in hiding them in its murky depths. Half-buried trees litter the channel at every turn. As with icebergs, only the emergent tips are apparent. Many more trunks and broken branches lurk just beneath the surface, some of them pointed, lance-like, in line with the current, ready to flip the boat or perhaps even split a hull. Two or three days of threading the log jams are required to reach our destination. The only landmarks are beaches, 105 of them at low water, and occasional high red banks marking the rare spots where the Manu touches the bounds of its broad meander plain.

The trip is a pleasant one in fine weather. Groups of macaws trade overhead, herons and kingfishers start up from the banks as we approach, and irate terns divebomb the boat as it passes their nesting beaches. One is constantly filled with the anticipation of seeing some rarer denizen of the region, a family of capybaras, a jaguar sunning on the bank, or even a startled native fleeing into the forest. This is perhaps what makes the mood of being at Manu so special, the subliminal excitement of knowing that one lives there by the tacit acquiescence of unseen and totally wild human beings. Now and then we come upon a trail of footprints, or a makeshift overnight lean-to on a beach. Once a pair of naked figures sprinting for cover. Nothing more. They are around us, and they know we are there. But we don't know who they are, where they live, or even what language they speak. So far it has been a peaceful coexistence, but there is no formal treaty, and one wonders what one would do if suddenly confronted by a group of them in the forest.

The long trip finally ends as we land on the 105th beach. A bold red-on-white sign perched atop the bank, an incongruous intrusion in the primeval scene, announces our arrival at the *Estación Biológica de Cocha Cashu.* A path leads into the forest, and we begin the sweaty job of packing more than a ton of supplies into the station itself, half a kilometer in from the river. After three days in the glare of reflected light, the cool, dark forest presents a welcome change. Massive trunks rise out of sight through tiers of foliage to broad crowns far overhead. One's perception of the scale of the forest is transformed. From midstream it appears diminished because there is no frame of reference by which to judge proportions. Inside, the perspective is entirely different. People are tiny and the trees are huge, domineering. After walking for less than ten minutes, a glow of light ahead presages our emergence into the station clearing, the only man-made break in the forest for many kilometers around. Though the opening is so small that many of the larger fringing trees would completely span it were they to fall, it provides light, air, and space, relief from the claustrophobia and dampness of the forest.

A split-level palm-thatched house on stilts fills one corner of the clearing, and just beyond it sunlight sparkles on the ruffled, pea-green water of a tree-rimmed lake. Seemingly contained by the curtains of vines that festoon its banks, its parallel margins curve gently out of view in the distance. This is Cocha Cashu, named half in Quechua and half in Spanish for the resemblance of its outline to that of the nut. From any room in the house one enjoys a view of the lake and full exposure to the welcome convectional breezes it generates in the middle part of every sunny day.

Somewhat pretentiously proclaimed as a biological station, the house was built in 1970 by biologists from the Universidad Nacional Agraria de la Molina (Lima, Peru) and the Frankfurt Zoological Society (West Germany). Since 1973 the facilities of the station have graciously been made available to us by the Peruvian Ministry of Agriculture, overseer of the park. The amenities are appreciable when one considers the alternative of camping in the forest: shelter, storage space, work benches, a large dining table, and refuge from mosquitoes. Beyond this, it is strictly a do-it-yourself proposition. There is no resident caretaker, no equipment, nor any organized form of transportation to or from the park. We provide all the logistics: boat, motors, tools, fuel, food supplies, etc. Everyone present at the station (usually eight to ten of us) shares in the chores. These are organized according to a rotating schedule—cooking, washing pots, hauling water from the lake, gathering and splitting firewood, and sweeping the house. Added to an active field program, chores, self-maintenance activities (bathing, washing clothes, etc.), and the social ritual of eating meals together fill out the days. The living, though far from luxurious, is thoroughly comfortable, and above all, blissfully removed from the cares of this troubled world.

I would here like to express my appreciation and gratitude to the many individuals and organizations that contributed to the making of this book.

First and foremost among these are my field companions. From the outset the project was conceived as a team effort. In no other way would it have been possible to accumulate 2,700 contact hours with our study troops in just a little more than one year. Once we had returned from the field, analyzing the data called for another four man-years of effort. Three people, Charles Janson, Debra Moskovits, and Grace Russell, contributed two or more years of their lives to the project, sharing with me the pleasures and frustrations of the field work and then continuing on at Princeton to organize and process the massive, at times almost overwhelming, quantity of notes and data. Many of the ideas set forth in the book sprang from the conversations we had during the year we spent together at Cocha Cashu. Charlie Janson made particularly important contributions toward designing the methodology employed in the observations, and in writing the many computer programs that were required to process the results. His penetrating insight into the major issues addressed by the project, and his impressive ability to articulate his views were a major source of stimulation and guidance throughout. Debbie Moskovits put the rest of us to shame in her capacity to recognize individual monkeys, to perceive subtleties of behavior that we simply did not notice, and in her incredible stamina for working long hours, both in the field and at the computer lab. Grace Russell provided the measure of good humor and understanding that held the rest of us together, and, despite no little inconvenience, remained steadfast in her dedication to seeing the project through to its final completion. Without her unflagging commitment, it is hard to imagine how the book could ever have been written. My indebtedness to all three is profound and I extend to them my deepest thanks. In the same breath I want also to thank Barbara Bell who joined us in the field for the latter half of the year. She cheerfully and capably caught up to the rest of us in her skill at making observations, even though it was her first experience in the forest.

There was one further individual who played an absolutely vital role in bringing the effort to fruition. This was Robin Foster, who provided the botanical expertise. Although good botany is indispensable to good primatology, we should have been nearly helpless were it not for the countless days Robin invested at the Field Museum identifying the hundreds of specimens in our collections. All the rest of us learned so much from Robin; merely thanking him for his altruism seems wholly inadequate.

For reliable and superbly executed assistance of an entirely different sort, we were all beholden to Klaus Wehr, my regular aide-de-camp since 1966. As guide, mechanic, cook, carpenter, boatman, and diplomat, Klaus is a master jack-of-all-trades who gets us into the field and keeps us there with a minimum time lost to extraneous details. He does the work of several ordinary people; we couldn't have done it so easily or so well without him.

In working in such an isolated spot as Cocha Cashu, one becomes unavoidably reliant on one's neighbors for hospitality and assistance in case of need, even though they may live dozens of kilometers away. Our neighbors were the park guards at Tayakome, six hours upstream, at Pakitza, two hours downstream, and the residents of the tiny settlement of Boca Manu, 200 km downstream. Among the park guards, Abel, Alciviares, Angel, Dino, Guillermo, Jacinto, and Jorge were especially friendly and helpful, and at Boca Manu we could always count on gracious hospitality at the home of Manuel Moreno.

Although the Manu Park has had official status since 1973, and has been under the protection of a staff of guards since 1970, it is still undeveloped as a tourist attraction. Permission to enter the park, and more importantly, to do research within its boundaries, requires the express written authorization of the Director General of Forestry and Wildlife within the Peruvian Ministry of Agriculture. The Director General at the time, Dr. Marc Duorojeanni R. and under him, the Director of Conservation, Ing. Carlos Ponce del Prado, are two of the prime movers behind Peru's outstanding conservation and parks program, and are among the people most directly responsible for the creation of the Manu Park. Their interest and encouragement were most crucial in our effort to reestablish Cocha Cashu as an active research station. We are greatly indebted to them for their broadmindedness and trust in helping to further the objectives of an unknown group of foreigners. At the local level we were given numerous forms of assistance by Adolfo Cuentas and Washington Galiano, the immediate administrators of the park. We wish to thank them both for their generous help on many occasions.

We would like also to thank the Cities Service Corporation for major assistance with our transportation needs during 1975–76 when it operated an exploration camp at Boca Manu.

Several individuals made substantial contributions to the preparation of the manuscript. Charles Janson wrote preliminary drafts of chapters six and eight that were extremely helpful in preparing the final text. Barbara DeLanoy persevered in her flawless typing through several drafts of the manuscript, often working after hours in an effort to meet deadlines. She also took care of countless details and served as my alter ego during many long absences in the field. For her unfailing warmth and cheer through all of these travails so efficiently performed, I shall remain forever grateful. Grace Russell and Debra Moskovits prepared many of the illustrations. For reading the manuscript with a fine-toothed comb and offering scores of valuable suggestions, I am especially indebted to my colleague Henry Horn. Tim Clutton-Brock, John Eisenberg, and an anonymous reviewer are also thanked for reading the entire manuscript and for making many constructive comments. Several others contributed by reading and criticizing selected chapters: Charles Janson, Debra Moskovits, Nancy Muckenhirn, Steve Schulman, and Harriet Smith. I would like to thank them as well.

Last but not least, it is a pleasure to acknowledge the generous financial support of the National Geographic Society (1974–75) and the National Science Foundation (DEB 76-09831 for 1976–78).

Princeton, New Jersey
January 4, 1983

Five New World Primates

1 Introduction

Primates are ideally suited for ecological study. Their size and diurnal habits put them comfortably within the range of human sensory abilities. Unlike birds, they can be followed and observed continuously throughout the daily activity period. But most importantly, primates gradually come to accept an observer as part of the landscape, or as just another participant in a mixed species troop. One gets to know each individual as a separate and distinct personality, and is privileged to observe the normal daily round of activity at close range without perturbing the course of events. The feeling of intimacy that one gains from this is certainly one of the greatest pleasures in studying primates. It is hard to think of another group of animals with such outstanding advantages.

But no matter how compelling these advantages may be, one does not simply walk into a new locality, even one as ideal as Cocha Cashu, and start observing monkeys. There are only a handful of places in the world where intact primate communities have been studied. Each one of them has required a major preliminary investment on the part of the investigators. At the very least, trails must be built and surveyed, and study troops must be habituated to the observers. Usually it is desirable to learn something about the population structure of the local primate community, the population densities of the various species, the size and composition of groups, the extent of home ranges, the amount of home range overlap, etc. Such background statistics are obtained by conducting extensive surveys employing protocols quite distinct from those used in detailed observations on habituated focal groups. For evaluating interspecific relationships it is important, in addition, to determine what the animals are eating. This calls for enlisting the collaboration of a skilled professional botanist, for the identification of plant specimens from most tropical regions requires a level of erudition that is utterly beyond the capacity of even a dedicated amateur. Finally, to interpret seasonal changes in diet and behavior, it is necessary to obtain data on the phenology of food resources. Obviously, not all these requirements can be met at once.

Our efforts began quite modestly in the summer of 1974. The station then offered little more than shelter; there were only the rudiments of a trail system and thus no ready access to much of the surrounding terrain. We had to spend the first weeks cutting trails on compass course, and were often surprised when the vegetation abruptly changed as we encountered swamps, canebreaks, and other features of the complex mosaic of habitats (Fig. 1.1). With the core of the trail system cut and surveyed, we spent the rest of the 1974 field season conducting a census of the primate community. As none of us had had any previous experience with primates, this also served to familiarize ourselves with the animals, and just as importantly, to familiarize the animals with us.

In this preliminary experience we were struck by the fact that among the eleven species present, there were great differences in troop size, social systems, activity patterns, and territorial behavior. There was nothing particularly unexpected in this, except that the first-hand experience of seeing it for oneself drove home the realization of how marked the differences between the species actually are. Another behavioral characteristic of some of the species also unavoidably attracted our attention. It was the tendency, very strongly developed, to join in mixed troops. This was especially interesting to me because at the time I was engaged in a study of mixed flocks of birds. Extrapolating from the avian flock systems with which I was familiar, there

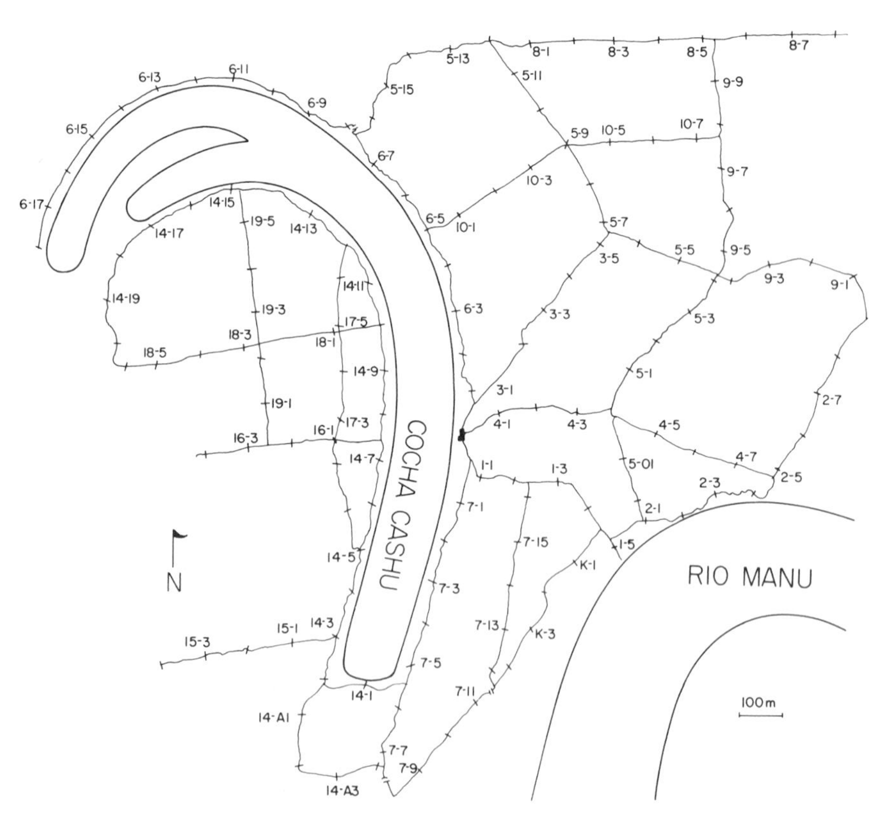

Figure 1.1 Map of the trail system at Cocha Cashu. Locations along the trails are marked with metal tags every 50 m (shown every 100 m on the map).

were some similarities in the behavior of the monkeys, but also some puzzling differences.

There were five primate species that participated in mixed troops. All of them fed on ripe fruit and devoted a considerable amount of time to searching for insects, and none of them was seen to eat leaves. They could thus all be classified as omnivores and be regarded as occupying the same trophic level. What was peculiar about them was that they displayed a wide range of group sizes, social systems, and territorial behaviors. Among birds, by and large, this is not the case. Species on the same trophic level that eat similar things tend to live in very much the same way. The monkeys violated this generality and therefore seemed paradoxical.

Two major issues thus forced themselves on our thinking. The first was, Why should a set of species which superficially seemed so similar in their feeding habits diverge so conspicuously in their social systems and use of space? The second was, What motivated the formation of mixed troops of primates when in some respects the circumstances were at odds with the better-understood avian situation? We were now convinced of the potential theoretical interest of a comparative study.

Returning in the summer of 1975, we set out for the first time to follow troops and to make systematic observations on the five species. We made many mistakes, including losing the focal group all too frequently, but we gained valuable experience. Most critically, we developed a protocol that permitted us to make maximum use of the limited time provided by an academic vacation. Following monkeys on a day-by-day basis is a very exhausting proposition. By working as a team (Janson, Moskovits, Russell, and Terborgh), we found that we could easily maintain contact with the study troops for longer than the five-day sample periods that are customary in primatology. The shifts gave everyone time to transcribe data, and to participate in other kinds of activities related to the project, such as mapping trails and servicing fruit traps. We also had time to ourselves, an important factor in maintaining morale through a long stint in the field. With some free time to look forward to each day, the routine never became oppressive. We were able to keep animals under observation continuously (except for brief interruptions when we lost troops) for practically the entire time of our residence at the station.

The following year I was given a sabbatical leave and at the same time received a grant from the National Science Foundation. These favorable circumstances provided the opportunity to carry out the main body of the study. In a 12-month period, starting in August 1976 and ending in August 1977, we logged over 2,700 contact hours with our study troops, for an average of about 540 hours per species. Here the advantage of working as a team is clear. It is common in studies conducted by single observers to limit troop following to 5 days per month. Over a year this mounts up to about 60 days of observation, a level we were able to match for several species. In place of short samples each month, we elected to conduct much longer samples (ca. 20 days) at less frequent intervals of about 3 months. This schedule permitted us to resolve the characteristic ranging patterns of each species, which in some cases are not discernible in five-day periods, as well as to compare the diets and ranging of each species within each of the four annual seasons: wet, dry, and the transitions between them.

While following the monkeys, we took four kinds of data: activity status of the troop, troop movements, use of resource trees, and timed samples of foraging activity. These provide basic information on time budgets, ranging behavior, diets, and foraging techniques. These ecological parameters were consciously emphasized to obtain the best possible picture of each species' interaction with its environment. Our approach necessarily relegated detail on behavioral interactions, vocalizations, etc. to a back-seat position. While this may prove frustrating to some readers, it is simply not possible to record everything about five species in a single year's study. Continuing research at Cocha Cashu will, over time, fill out the behavioral picture for the five species included in this study, and for others as well.

The comparative approach is followed throughout. Conclusions are drawn deductively or inferentially and are based on the analysis of differences between the species. The adaptive meaning of what a single species does (in the absence of a comparative context to provide a frame of reference) is very difficult to interpret. The common practice in primatology of studying one species at a time has provided a great deal of valuable information on many species, but such studies have often failed to generate incisive questions, much less answer them. The comparative approach provides the necessary perspective for framing the salient questions. For this reason I have organized the book not as a series of chapters on each of the species, but as a set of topical discussions in which differences between the species are analyzed and interpreted.

The presentation consists of eleven chapters. Following the present introduction are two additional preliminary chapters that describe, respectively, the vegetation and climate of the study site, and the primate community. The text then continues with a series of chapters that lay out the principal results. Activity budgets are taken up first because a consideration of how animals spend time provides insights into the priorities in their lives. The next two chapters, numbers five and six, offer data on the diets of the five species as represented, respectively, by the amount of time spent feeding on various kinds of plant materials, and by timed sequences of the actions employed in foraging for animal food. The seventh chapter presents data on ranging behavior. Large differences between the species in their responses to a seasonally fluctuating food supply are brought out particularly strongly in this chapter, though the theme is one that appears in nearly every section of the book. In Chapter 8 the problem of mixed species associations is addressed in the light of the two main types of associations at Cocha Cashu, those between *Cebus* spp. and *Saimiri*, and between the two species of *Saguinus*. The principal adaptational variables involved in structuring ecological relationships in the entire Cocha Cashu primate community are the subject of Chapter 9. Our findings are compared with those of other studies of sympatric primates in other parts of the world. In the tenth chapter I attempt to synthesize what has been learned previously, emphasizing results that bear on observed differences among the five species in the use of space, territorial behavior, and mean group size. The eleventh and final chapter is a brief epilog that offers suggestions on how tropical forests might be managed in the future to provide resources for both humans and primates.

2 The Study Site: Its Climate and Vegetation

Here I describe the environmental setting at Cocha Cashu. The chapter opens with a cursory account of the several types of vegetation that enter into the local habitat mosaic. Most of these are seral stages in successional sequences initiated by changes in the course of the river. Climatic data are presented next. Even though complete records exist for only a single year, there is no doubt that the region typically experiences a sharp alternation between wet and dry periods. This has far reaching consequences in driving the annual cycle of flowering and fruiting in the vegetation. In the latter part of the chapter it is demonstrated that the productivity of fruit by the forest follows a seasonal cycle of large amplitude, a fact that has profound implications for animals that depend on fruit as their primary food resource.

Habitats at Cocha Cashu

Riparian Succession

Cocha Cashu and all the terrain that lies around it on every side are creations of the Rio Manu. The sinuous meanderings of the river transform the landscape with a nearly imperceptible, though relentless, regularity. At each bend the forest is undermined and carried away at a rate of 5–20 m a year, while on the opposite, inside shore, new land is laid down in the form of broad beaches. Later, as the beaches are invaded by plants, they become incorporated into an incremental series of levee-backwater units. Thus, to walk into the rank vegetation bordering a beach is to walk backward in time, toward progressively older terrain, and progressively older vegetation.

Fronting the beaches are even-aged stands of *Tessaria* sp. (Compositae), a fast growing willow-like tree that springs up each year from seed at the beginning of the dry season. After two or three years, the *Tessaria* is invaded by heavy 10 m tall stands of caña brava (*Gynerium sagittum*, Gramineae), which advance up to a dozen meters a year by means of vigorous underground runners. At this stage the *Tessaria* declines rapidly, half or more of the trees dying each year, so that few individuals see even as much as a decade. The caña phase is more prolonged. It is during this period that many tree species become established, including the ones which dominate later stages in the succession. The fastest growing of these is a species of *Cecropia* (Moraceae), which, in about 5 years, succeeds in overtopping the caña. For a brief interval *Cecropia* forms a nearly closed canopy over the caña, but *Cecropia*, like *Tessaria*, is short-lived, and gaps soon appear in the canopy. These are promptly filled by a number of other species: *Erythrina* spp. (Leguminosae), *Guarea* sp. (Meliaceae), *Casearia* sp. (Flacourtiaceae), *Sapium* sp. (Euphorbiaceae), *Ficus insipida* (Moraceae), and *Cedrela odorata* (Meliaceae). The latter two of these outgrow the rest, eventually forming a closed canopy at a much higher level than the *Cecropia* (ca. 40 m vs. 15–18 m). Unlike the preceding stages, the *Ficus-Cedrela* forest is long lived, with individual trees commonly attaining diameters of 1–1½ m, and ages well in excess of 100 years. The understory at this stage is a 3–4 m tall jungle of "*platanillos*," broad-leafed monocotyledonous plants in the banana (*Heliconia*-Musaceae), ginger (*Costus, Renealmia*-Zingiberaceae), and marantad families (*Calathea, Monotagma*-Marantaceae). These are more shade tolerant than the caña brava and invade subsequent to the *Cecropia* phase as the over-topped caña goes into decline.

Large areas in the Manu basin adjacent to the river are occupied by successional vegetation. The interiors of meander loops are filled with it. Because of the great regularity with which plant species replace one another in the canopy, and because each stage reaches a greater absolute height than the previous one, the successional vegetation is strikingly zoned. As viewed from the vantage of a passing boat, one sees an orderly sequence of stands—*Tessaria, Gynerium, Cecropia*, mixed forest, and *Ficus-Cedrela*—each forming a nearly discrete band of greater or lesser width running parallel to the adjacent beach. Of these, the *Ficus-Cedrela* association occupies by far the largest area by virtue of its vastly greater longevity. Eventually, however, the *Ficus* and *Cedrela* trees begin to die off, one by one, and are replaced by an entirely different set of species which form a forest of considerably greater diversity. The succession continues, in other words, to even more advanced stages, but the details have not been studied.

Successional vegetation along the Rio Manu is exposed to periodic inundation during the rainy season. Flooding results in the deposition of a fresh layer of silt each year, rejuvenating the already high fertility of the alluvium, and gradually raising the elevation of the terrain to equal that of the highest water levels. Tree growth is extremely rapid. Young specimens of *Ficus insipida, Ochroma pyramidalis* and other species may increase in circumference as much as 100–200 mm, or up to 30%, in a single year. Trees of similar size in climax forest rarely achieve growth rates even one-tenth as great.

Whatever the reasons for this contrast, it is evident that successional stands are highly productive. This fact is widely recognized by immigrant settlers and indigenous inhabitants alike, as their agricultural efforts are typically concentrated in the upper levels of the flood zone. The high productivity of this zone, coupled with the breathtaking value of its *Cedrela* timber, has encouraged an intensive selective exploitation of *Ficus-Cedrela* forest wherever it occurs in the Amazonian headwaters region. Little of it remains in natural condition outside the Manu park.

Lacustrine Succession

Over many millennia the Rio Manu has carved a 6–8 km wide meander belt into the local topography. Wherever it contacts the fringing hills, the river cuts into them, creating steep banks, and depositing fresh silt in place of the eroded sediments. By this process, the meander belt is gradually widened, but judging from the paucity of such contact points, the rate of widening today is extremely slow. The fringing hills, barely 20–30 m above the present river level, are themselves composed of sandy alluvium, probably the remnants of a previous outwash plain.

Wherever one goes in the meander belt, telltale signs of the river's activities are apparent. Slightly raised ridges, representing ancient levees, and adjoining depressions, representing former backwaters, are the most common indications. More conspicuous are sections of river channel, long isolated from the main stream. These support a vegetational succession quite distinct from the one described above. The initial stage is that of an oxbow lake, formed when a meander loop is suddenly short-cut in a flood and isolated from the newly created channel. The subsequent fate of the oxbow lake is strongly dependent on the extent to which its water level is coupled with that of the river. The outlet channel of some lakes is cut deeply enough to provide a close connection. Such lakes rapidly fill up with sediment, since each incursion of river water carries with it a new load of silt. Lakes of this type have a relatively brief existence, less than the lifespans of the first generation *Ficus* and *Cedrela* trees that line their margins. Other lakes, and Cashu is one of them, are relatively uncoupled from the river. Only the highest floods deliver a charge of sediment, and then a relatively small one. Such lakes are deeper (Cashu is 2 m), and endure at least for centuries, as judged from the successional state of the vegetation on their shores.

Lake-bed succession begins with a covering of floating aquatics (*Pistia stratiotes*: Araceae), which reach a cyclic peak of cover and luxuriance during the latter part of the flood season, presumably in response to the fresh supply of nutrients contained in the silty river water. Cashu, being relatively uncoupled, has virtually no floating aquatics, only planktonic algae. As sedimentation proceeds, certain grasses, sedges, and other semiaquatic plants begin to encroach from the shores, at first forming plugs of vegetation and floating islands, and later crowding out the open water altogether. In turn, the grass stage is invaded by shrubs, especially *Annona tessmannii* (Annonaceae), and somewhat later by *Ficus trigona*. The long-lived subclimax of lake-bed succession is a *Ficus trigona* swamp, the analog of the *Ficus insipida-Cedrela* association in riparian succession.

Occupying elongate depressions in the forest, *Ficus trigona* swamps present a strange and fantastic scene. The horizontal trunks sprawl, seemingly endlessly, in all directions, being supported by festoons of prop roots. The trees convey no sense of discreteness; it is as if the whole swamp were enveloped in the ramifying tentacles of a sinister creation of the vegetable kingdom, a real life embodiment of the imagination of fairy tale and science fiction writers. Much of the year the swamps are knee or waist-deep in water. It is only in the dry season that one can comfortably enjoy their exotic flavor. *Ficus trigona* is an especially important tree in the economy of the forest, because the fruits are favored by a wide range of primates from tamarins to spider monkeys, by other diurnal mammals (e.g., coatis), and by many birds as well. In fruiting season the swamps are alive with activity; at other times they are virtually empty.

High Ground Forest

High ground forest occupies much of the greater part of the meander belt, and constitutes the matrix within which the river, its lakes, and the various types of successional vegetation just described are set. By the Holdridge classification system (Holdridge 1967), it is tropical moist forest. Flying over it, the impression one gains is of endless uniformity, a textured but monotonous carpet of green that extends unbroken to the horizon. Yet the more one becomes familiar with its subtle internal details, the more one is impressed with the distinctness of each small section. Slight variations in the species composition of the canopy, in the frequency of treefalls, in the prevalence of palms, in the density of lianes, and in the character of the understory all contribute to a kaleidoscopic variability. The factors which contribute to this are undoubtedly many—successional age of the site, drainage, water table, exposure to inundation, frequency of treefalls, time since the last major disturbance, etc. In some instances the correlates can be deduced. For example, certain sectors of the forest seem to be underlain by a layer of hardpan. Impeded drainage results in standing water during much of the rainy season, with the consequence that treefalls are exceptionally frequent. Openings develop which rapidly become choked with 4 m tall stands of *Heliconia episcopalis* (Musaceae), a condition that greatly retards the regrowth of trees.

The steady state is a large-scale dynamic patchwork of copses, scattered trees, and platanillo (*Heliconia*) openings.

On well-drained sites the forest develops impressive proportions, with emergents attaining heights of 50 or even 60 m. Among these, *Dipteryx micrantha* (Leguminosae) is preeminent, both in its abundance (ca. 1 per ha) and in its immense stature. Beneath the crowns of emergents, one finds superimposed an average of four additional layers. Thus, in its vertical organization, the forest conforms to the five-tiered archetype propounded by Richards (1952). Typical crowns of successively lower strata are smaller and smaller, a matter that is of considerable consequence to primates. A normal tree in the subemergent layer may have a height of 30 or 35 m and a spread of 10–15 m. This is the size of a large oak or ash in a mature North American forest. Yet as many as six to eight of these comfortably nestle under the outstretched limbs of a single *Ceiba, Dipteryx*, or *Ficus perforata*. It is the emergents that endow the tropical forest with its undeniable majesty.

In the middle tiers of the forest there is rampant confusion, or so it seems. Crowns of many sizes and shapes crowd one another both vertically and horizontally. Lianes arch through the branches, forming interdigitating networks that unite large blocks of trees in a common web of entanglement. Palms of several kinds are abundant, contributing in a major way to both the appearance and the economy of the forest. Some (*Iriartia, Socratea*) bolt upwards on stilt roots to fill gaps in the canopy, while others (*Astrocaryum, Scheelea*) grow imperceptibly in their tolerance of conditions in the shade. There is hardly a species of palm, of more than twenty in the Manu region, that is not avidly used by some animal or another. The forest without them would be a different place.

There is less to be said of the understory because the plants are individually small and mostly incapable of supporting the weight of monkeys. Many of them produce red or dark purple berries that are dispersed by birds (e.g., many Rubiaceae, Melastomaceae, *Neea* spp.-Nyctaginaceae, *Ardisia* spp.-Myrsinaceae). The ground layer is singularly depauperate in plant species. On well-drained soil it consists of a nearly monotypic lawn of ferns (*Tectaria incisa* var. *vivipara*-Polypodiaceae). Wherever there is standing water in the rainy season, the ferns are replaced by a low species of *Heliconia*.

To describe the many facies of the high ground forest at any further length would be to indulge in a welter of botanical detail that would go beyond the likely interests of our readers. Suffice it to say that the species diversity is very high. Our plant collections to date have included well over 1000 species and are far from complete. The greater part of this intimidating botanical richness is found in the high ground forest. Successional vegetation is rich by temperate standards but does not approach the variety contained in the well-drained climax forest.

A final point about the vegetation in the study area is that it is a complex mosaic of habitats (Fig. 2.1). There are sizeable areas of both riparian and lacustrine successional vegetation, exposed edges along the lake and riverfronts, and tall forest of varying age, successional status, and exposure to inundation. The significance to primates of the floristic differences between habitats is not so much that resources are produced in different qualities or quantities, but that they are produced on differing phenological schedules, as shall be seen in a later section of this chapter.

The Climate at Cocha Cashu

Our knowledge of the climatic conditions at the study site comes from a year of records taken between August 1976 and August 1977, and from less

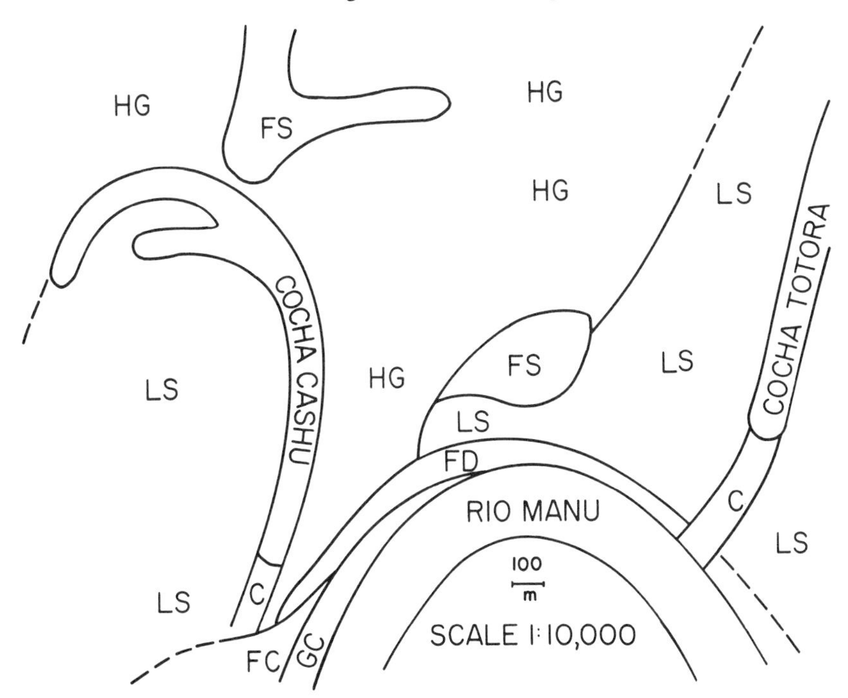

Figure 2.1 Distribution of habitats in the Cocha Cashu study area. Key: GC = riparian canebreaks of *Gynerium sagittum* overtopped by a canopy of *Cecropia leucophaia*; C = late successional forests dominated by *Cedrela odorata*; FC = successional forest dominated by *Ficus insipida* and *Cedrela odorata*; FD = flood-disturbed forest of varied structure and composition; LS = late successional forest of high species diversity (mostly subject to seasonal inundation); FS = fig swamp dominated by *Ficus trigona*; HG = mature high ground forest.

systematic observations during several shorter periods of residence (Table 2.1). Situated at 11°51′S, 71°19′W, the station is on the fringes of the equatorial zone, and consequently experiences a marked alternation of wet and dry seasons (Fig. 2.2). Total rainfall recorded in the year was 208 cm, of which 180 cm (87%) fell in the seven wet months (October–April), and 28 cm (13%) in the five dry months (May–September). Although we have no way of knowing whether this was a typical year, the rainfall regime closely resembles that at the Smithsonian research station on Barro Colorado Island, Panama (Leigh and Smythe 1978). The long, and at times severe, annual dry period results in a marked lowering of the water table, as indicated by a progressive seasonal drop in lake level. Moisture stress in the forest is manifested by a considerable degree of deciduousness in the canopy, and by a

Table 2.1
Summary of climatic records at Cocha Cashu, September 1976 through August 1977.

	Rainfall				*Temperature °C*			
		No. days with						
Month	*Precipitation (cm)*	*No rain*	*Trace*[a]	*Rain*	*Mean maximum*	*Mean minimum*	*Monthly mean*[b]	*Cloudiness mean*[c]
September	5	18	7	5	27.7	20.4	24.1	0.98
October	27	20	4	7	**28.9**	21.9	**25.5**	1.21
November	22	18	5	7	28.6	22.1	24.0	1.72
December	19	12	5	14	27.0	22.3	24.7	1.68
January	29	12	6	13	27.2	22.6	24.9	1.71
February	18	8	4	16	27.5	22.4	25.1	1.85
March	39	11	1	19	27.0	21.7	23.9	1.44
April	27	17	1	12	25.1	21.7	23.5	**1.89**
May	9	22	2	7	24.3	20.4	22.6	1.38
June	5	23	2	5	24.5	**20.0**	**22.2**	0.65
July	2	22	5	4	26.4	21.2	24.1	1.14
August[d]	8	10	2	1	28.0	20.9	24.4	**0.48**

NOTE: Boldface entries indicate maximum and/or minimum values for column.
[a] Amounts <1 mm were recorded as trace.
[b] Monthly means computed as the mean of daily midpoints.
[c] Cloudiness index derived from assigning values to each half-day as follows: clear = 0; partly cloudy = 1; cloudy = 2; overcast = 3.
[d] Incomplete month; figures based on 13 days of records.

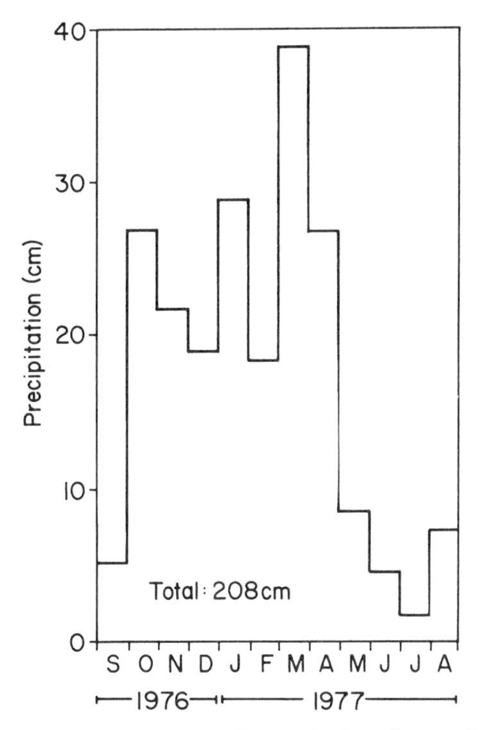

Figure 2.2 Monthly precipitation at Cocha Cashu: September 1976 through August 1977.

daily cycle of wilting in understory herbs, especially those in openings that are exposed to midday sun. Many canopy trees show a partial response by reducing their foliage volumes and then waiting until after the first heavy rains before sprouting new leaves.

The severity and timing of the dry season can vary appreciably between years. In 1974 and again in 1978, steady rains continued into July, with a consequent lag in the onset of leaf fall, and probable reduction in the duration of deciduousness in the canopy. In contrast, rains ceased abruptly in early May of 1977, leading to early and pronounced leaf drop. In 1976 the dry season ended with several heavy downpours in October that quickly replenished the water table, filling numerous depressions in the forest. But October of 1979 was as dry as the two preceding months. Such pronounced year-to-year variation in the onset and termination of the dry season could have important consequences for the fruiting phenology of tree species important to primates and other consumers (Chivers and Raemaekers 1980). This is an interesting issue which can only be addressed with long-term records that are currently unavailable.

Although the flora and fauna of the Manu basin are typical of the Amazonian lowlands, the elevation of the study site (ca. 400 m) is sufficient to moderate the oppressive heat that one experiences further downstream (Fig. 2.3). Temperatures of 30° (86°F) or more occur only during the months of August, September, and October. In over 600 days at the station, the highest temperature on record is 33° (91°F). Normally, the daily excursion of 4–6° shuttles between the lower and upper 20s, a quite agreeable range. The only dramatic departures from the normal pattern are connected with major storm fronts accompanying the southern winter. Every year several of these sweep up the continent, sometimes affecting abrupt and drastic drops in temperature as far north as Iquitos (lat. 3°S). Such frontal systems can arrive any time from May until early September, but are most frequent and pronounced in June and July. The most severe of these storms in our experience came in July 1975, when a low of 8° (46°F) was reached for three nights in succession. This was the storm that froze most of the coffee plants in Brazil, and which resulted in widespread damage to natural vegetation as far north as the state

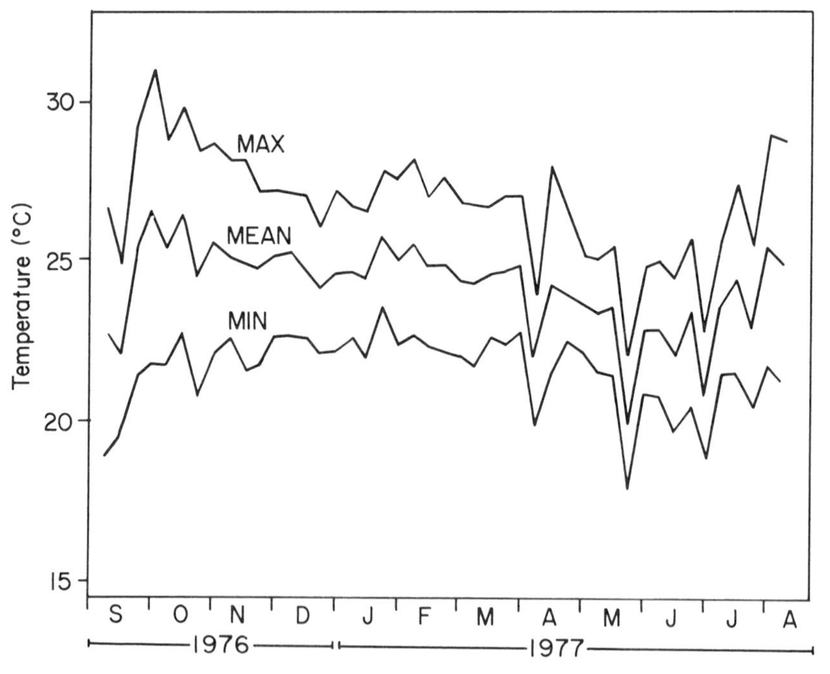

Figure 2.3 Temperature regime at Cocha Cashu between September 1976 and August 1977. The values plotted are weekly means for daily maximum, mean, and minimum temperatures. Daily mean temperatures were estimated as the midpoints between daily maxima and minima.

of São Paulo (Silberbauer-Gottsberger et al. 1977). Of course, this was an extraordinary, once-in-a-century event. In a more typical year, the minimum temperature is in the range of 14–16° (57–61°F).

Fruiting Phenology

Since fruits constitute the principal source of calories for most primates, including all the ones considered here, it was important to have some measure of fruit availability as it varies seasonally. Following Smythe (1970) and Foster (1973), we set traps in the forest to capture fruit falling from the canopy.

The traps were set 20 m apart and 1 m to the side of convenient trails. They consisted of light plastic trash bags suspended about 0.4 m above the ground from wire hoops 1 m in circumference. Each trap thus had a collecting area of 0.08 m^2. Rainwater was allowed to escape through small holes cut in the bottom corners of each bag. Although the total area of the traps was not large, adjacent traps were far enough apart in most cases to provide independent samples. Collections of fruit and seeds entering the traps were conducted on a regular biweekly schedule. The contents were sorted by species and fruit type (drupe, berry, capsule, etc.) and weighed while fresh. By taking three sample running means, sample-to-sample variation was smoothed out somewhat to provide a better representation of long-term seasonal trends.

A group of 50 traps was arrayed in each of three more or less distinct habitats which shall be called forest, levee, and river. The forest traps were in old growth, well away from either the river or lake margin. A number of them were located in regenerating treefall openings, or in shallow depressions that collect water in the rainy season. These microhabitats, as well as tall forest on well-drained ground, were sampled in more or less the proportions in which they occur in the study area. The levee traps were strung along the raised eastern margin of the lake, mostly within 10–20 m of the water's edge. Finally, the river traps were set along a trail that parallels the riverfront for 400 m (20 traps), and then turns inland through seasonally inundated, late successional riparian forest. Tree species composition in the sectors covered by the forest and levee traps is similar, and markedly distinct from that in the riparian zone sampled by the river traps.

Before discussing the results, it should be pointed out that fruit traps measure a residual quantity: total fruit production by the canopy minus the amount eaten by arboreal frugivores, including insects. So far as we are aware, gross fruit production has never been measured in a tropical forest. Fruitfall data can be misleading because they overestimate the seasonal variation in fruit availability. This is because arboreal frugivores can be expected to eat a greater proportion of the available crop when fruit is relatively scarce than when it is superabundant. How much is eaten is not known, but it may be a substantial portion of the gross production in all but the peak months. This much can be surmised from the presumed metabolic demand of the frugivore community, which at Cocha Cashu constitutes a biomass of around 20 kg/ha (see Chapter 3). About half of this biomass is made up of arboreal frugivores. Very crudely, then, if the animals in question eat an average of 20% of their weight per day in fruit (e.g., Leigh and Smythe 1978; Nagy and Milton 1979), they will consume 2 kg × 30 days, or 60 kg/ha per month. This is an amount that exceeds the quantity that fell into the traps during the low period of the year, notwithstanding the fact that many of the fruits captured by the traps were of types that are not eaten by frugivores. It is clear then, even though the numbers are very rough, that arboreal frugivores consume a significant fraction of the fruit produced by the forest. The residual (uneaten) fruit that collects in traps is thus strongly biased against edible fruits, especially at times of low abundance. The error is one that affects the perceived magnitude of seasonal fluctuations in abundance, but which should not distort the positions of the peaks and valleys in the seasonal pattern.

Being forearmed with this proviso, we can now proceed to inspect the results (Table 2.2 and Figs. 2.4 and 2.5). Seasonal fluctuation is very pronounced. In all three habitats there appear to be two peaks in fruit production, one early in the rainy season (November–December) and one later (February), about two months prior to the onset of the dry season. Minimum fruit drop was in May (June in the river traps), which in 1977 was the first dry month. The pattern differs somewhat from the one recorded by Smythe (1970) on Barro Colorado Island, Panama. He found a major peak in June (corresponding to the December peak at Cocha Cashu), but no secondary peak later in the rainy season. Instead, there was a minor peak in February (dry season), for which we find no counterpart in our data. However, the two sets of results agree quite closely in pointing to the early part of the dry season as the period of minimum fruit availability (December–January, Panama; May–June, Peru). More surprisingly, there is also good agreement at the quantitative level in the total annual fruit drop in the two localities: ca. 2180 kg/ha (Panama) and 1990 kg/ha (Peru: mean of the three sets of traps).

An important concomitant of seasonal scarcity of fruit at Cocha Cashu, and presumably elsewhere in the tropics, is that fruiting trees are farther apart. This is implied, albeit somewhat indirectly, by a decrease in the number of traps capturing fruit each biweekly sampling period (Fig. 2.5). The time course of this measure of the frequency of fruiting trees closely parallels that of gross fruit drop, even to the extent of revealing a bimodal peak during the period of high production. Thus, as also seems reasonable on *a priori* grounds, abundance and scarcity are primarily accounted for by the number of trees

Table 2.2
Weight of fruit falling each month in three habitats at Cocha Cashu.

Month	*Habitat*			*Total*
	Forest	*Levee*	*River*	
September	115	154	106	375
October	68	147	185	400
November	107	175	329	611
December	**236**	**383**	242	861
January	72	182	187	441
February	130	287	**1021**[a]	**1438**
March	119	239	276	634
April	76	186	128	390
May	**29**	**8**	69	**106**
June	53	149	**37**	239
July	65	187	39	291
August	75	66	44	185
Total for year	1145	2163	2663	1990[b]

NOTES: Units are kg per ha. Divide numbers by 10 to convert to metric tons per km^2. Maximum and minimum values in each column in boldface.

[a] High monthly total due principally to a large fig that dropped fruit into two adjacent traps.

[b] Mean of the three annual totals.

fruiting, rather than by the amount being dropped per tree. To the degree that this is true, it follows that the amplitude of seasonal fluctuations in fruit availability is more faithfully mirrored in the numbers-of-traps-with-fruit statistic than it is by the gross quantity collected, because it is probably less sensitive to the activities of arboreal frugivores. The bottom of the May–June trough in these data represents roughly a 10-fold decline from the maximum level in the number of traps capturing fruit. Fruiting trees are not only fewer, but much farther between, a condition that carries major implications for the ranging behavior of frugivores, as we shall see in Chapter 7.

Fruitfall by Type

Marked seasonal patterns are evident, not only in the amount of fruit produced by the forest but in the types of fruit as well (Figs. 2.6 and 2.7). The

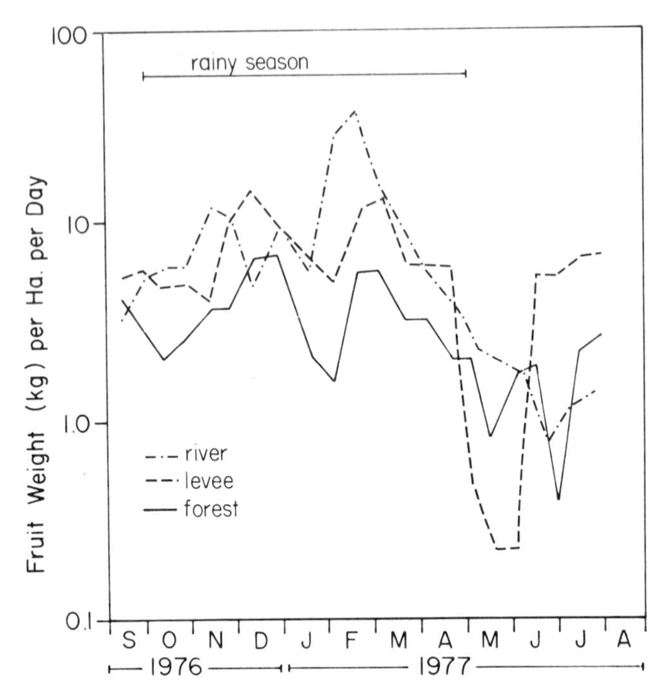

Figure 2.4 Fresh weight of fruit falling from the forest canopy at Cocha Cashu. Values represent running means for three biweekly collections from 50 traps.

classification of fruits into three categories is based more on our immediate interest in their potential as food for primates than it is on botanical considerations. One category (crosshatched in the figures) includes all fleshy fruits, mostly drupes and berries, the great majority of which are animal dispersed, and at least potentially of importance to primates. The second category (simple hatching) consists of dry fruits (pods, capsules, and samaras) whose seeds are primarily wind dispersed. And the third category is composed of hard fruits (nuts, achenes, etc.) and detached seeds that had been dropped or defecated from the canopy. Fruits belonging to the latter category are consumed primarily by specialized terrestrial mast feeders—rodents, peccaries, tinamous—although some are harvested before they fall by *Cebus* monkeys and macaws.

The broad features of the pattern of fruitfall with respect to type are in good agreement with previous measurements undertaken in Costa Rica (Frankie et al. 1974) and Panama (Smythe 1970; Croat 1975). In both major habitat types, interior high ground forest and seasonally inundated riverine forest, there is a pronounced release of wind-borne fruits in the early half of the

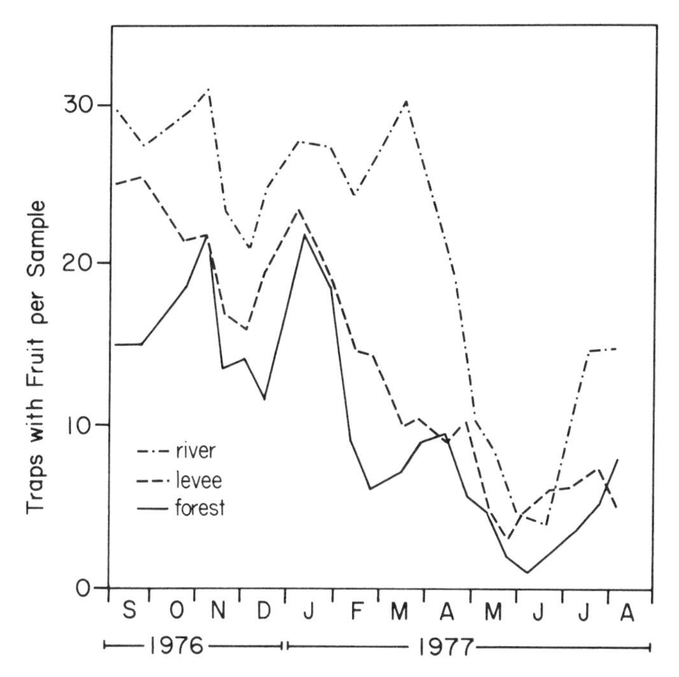

Figure 2.5 Number of traps out of 50 that contained fruit each biweekly sampling period. Values represented are running means for three successive collections.

rainy season. Both the absolute and relative abundances of wind-dispersed fruits are decidedly greater in the riverine forest, and the period over which they are dropped is more extended. Nevertheless, the release of such fruits in all habitats goes virtually to zero at the beginning of the dry season. This time is no doubt the most stressful one of the year for freshly germinated seedlings, and would be especially hostile to the tiny sprouts that arise from most wind-carried seeds. The relative prominence of wind-dispersed species in the riverine forest can perhaps be explained by the occurrence of greater wind velocities and updrafts along the exposed forest edge, and by the presence of numerous disturbed sites such as banks, bars, and washouts that are annually available to invasion.

Soft fruits (drupes and berries) follow the same overall pattern as fruits belonging to the other categories in being produced in the greatest variety in the early and middle parts of the rainy season. However, the fruiting seasons of plants adapted for animal dispersal are better staggered, so that some species are available at all times. This observation has been made previously in Panama (Hladik and Hladik 1969; Smythe 1970) and in Costa Rica (Frankie

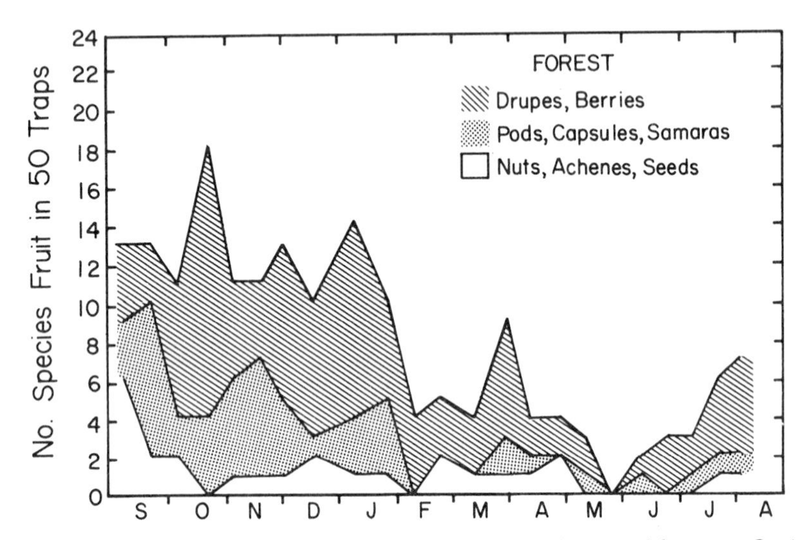

Figure 2.6 Distribution of fruit types falling from high ground forest at Cocha Cashu. Explanation in text.

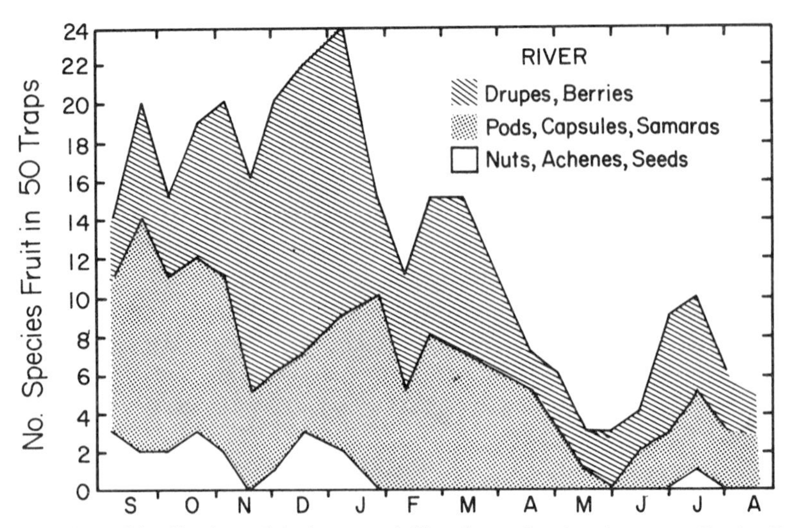

Figure 2.7 Distribution of fruit types falling from riverine forest at Cocha Cashu. Explanation in text.

et al. 1974 at La Selva), and thus appears to be quite general for the Neotropics. Much of the soft fruit that ripens at Cocha Cashu between April and August is in the form of figs, of which there are more than twenty species.

Phenology in Minor Habitats

The final point to be made about the phenology of fruiting at Cocha Cashu is that there is a definite sequence in which different habitats reach their fruiting peaks. Only the two most prevalent habitats were monitored with fruit traps. Several of the minor ones show quite divergent behavior.

Ficus trigona swamps, for example, begin to fruit in August and continue into October. They are heavily utilized at this season, and practically ignored at others. By the time the *F. trigona* in the swamps begins to decline, there is abundant fruit of many kinds in the high ground forest, a situation that persists (or did in the year of our study) into February. By then, large quantities of *Inga* pods (especially, *I. marginata* and *I. mathewsiana*, Leguminosae) are ripe all along the margin of the lake, and to a lesser degree along the riverfront. These are intensively exploited, particularly by *Cebus* spp., for two or three weeks until the trees are practically bare. The focus of activity then shifts to the successional forest in the sedimented river channel that is the prolongation of the lake. Here there is an abundance of *Xylopia* (Annonaceae) and other fruits which holds until mid-March, whereupon there ensues a mass movement of primates into the canebrakes bordering the river. Although little used at other seasons, the canebrakes offer a dual attraction just at the time when resources are beginning to decline precipitously in the forest. One of these, surprisingly, is insects. In March and April the canebrakes are flooded, and consequently, nearly the entire leaf-litter arthropod fauna must take refuge on the stems of plants, where the hapless creatures are exceptionally vulnerable to low-foraging birds and squirrel monkeys. The other attraction is a great abundance of ripening *Cecropia* (Moraceae) catkins, supplemented by quantities of *Ficus insipida* and *Casearia* sp. (Flacourtiaceae) fruits. The supply is bountiful, due to the broad expanse of homogeneous vegetation, so that it suffices for a major part of the primate community for well over a month.

Because the evidence for a sequence of fruiting peaks in minor habitats is anecdotal, and based on the behavior of frugivores rather than direct measurement, it could be argued that the sequence is one of preference by the animals, rather than of ripening. There are, however, some objective reasons for thinking it is real. One is the direct observation that *Xylopia* (which dehisces to expose bright arils inside) and *Cecropia* did in fact mature late in the season after many other common species had been depleted. Another is that the amount of fruit falling into the forest and river traps was in a steady and precipitous decline during March and April. Minor habitats such as the successional lake bed and riparian vegetation characteristically possess low plant species diversity. Some are nearly pure stands of a single dominant species (e.g., *F. trigona, F. insipida, Cecropia* sp.). During much of the year these habitats are visited by primates only incidentally as they transit from one part of their home range to another. The fruiting behavior of minor habitats is perhaps designed to attract animal dispersers from other habitats. To be most effective, the prevalent species should fruit in synchrony and at a time when other habitats are decidedly less attractive. Otherwise, the level of frugivore interest would not be as high as it could be, and perhaps worse, dispersers might remain for brief periods only and defecate the seeds in inappropriate sites (Howe 1979). Thus, from the point of view of the plants, maximum attraction and retention of dispersal agents is achieved by having a concentrated, yet isolated, fruiting period. Beyond satisfying these criteria, the actual time of fruiting may be subject to a variety of additional constraints of a physiological kind operating on the parent plant and/or its seedlings.

From the limited perspective of a single year's sojourn at the study site, it is nonetheless clear that the subject of fruiting phenology is one of enormous importance and complexity. We have reviewed only the most obvious patterns, and these at the aggregate level of whole habitats. Underlying each pattern is a series of tradeoffs involving optimization of physiological effort on the part of the plants and the many exigencies of dispersal. These tradeoffs regulate not only the broad outline of the seasonal march of fruit availability, but no doubt adjust the steady state blend of fruit types as well. Out of the several concurrent patterns, two broad trends emerge as the most important to monkeys. One is the regular annual alternation between a period of relative abundance (September–April) and a period of relative scarcity (May–August). The other is the staggering of fruiting peaks between major and minor habitats, and between minor habitats *inter se*. The first of these appears to be a general occurrence over much of the Neotropics. The second can only pertain to a complex mosaic environment, but may prove to be widespread once it is better studied. In any case, both patterns appear to be crucial to molding the lives of animals at Cocha Cashu.

3 The Primate Community at Cocha Cashu

Primates in the Community of Frugivores

Throughout most of this work we shall focus with myopic concentration on the lives of five species of primates. This is an inevitable and unavoidable weakness of an intensive study. In full realization of this, it will be helpful to place our subjects in perspective before narrowing the frame of reference.

An untouched tropical forest standing on rich alluvial soil, such as the one at Cocha Cashu, veritably abounds in animal life. To one accustomed to the temperate milieu, the contrast is astonishing. The most conspicuous difference is in the prominence of mammals. There are scores of species, many of them large and numerous, in comparison with a paltry complement of mice, deer, and squirrels at higher latitudes.

What is to account for this extraordinary difference? It is due largely to the year-round availability of fruit, produced on a scale of tons per hectare-year (see Chapter 2). Of the total frugivorous biomass, primates contribute a minor fraction; among all the primates, the five we studied constitute again a minor fraction. We are looking then at a small part of a very much larger universe. This fact should always be kept in mind, even after I have seemingly lost sight of it myself.

I shall now attempt to frame the perspective in quantitative terms, although the numbers in some cases will be only crude approximations or educated guesswork. Discounting insects and fungi, which destroy large quantities of fruit before it ripens and after it falls, the major classes of frugivores are birds and mammals (Table 3.1). In temperate forests birds are the more important (Baird 1980), but in the tropical forest mammals predominate by a wide margin (Gautier-Hion et al. 1980). Our estimates are derived from a number of sources, some published and others not. Groups of frugivores that have been censused at Cocha Cashu are: primates (Janson 1975), peccaries (Kiltie and Terborgh 1983), and nocturnal nonflying mammals (L. H. Emmons, unpbl.). Data on birds come from experienced guesswork and some census results of Rolando Gutierrez on Cracids (curassows and guans). The numbers may be off by as much as a factor of two, but even so, the error induced in the total frugivore biomass is relatively minor.

Two further groups of species remain to complete the picture: diurnal

Table 3.1
Biomass density of mammalian and avian frugivores at Cocha Cashu.

Taxon	*Approx. no. individuals per km²*	*Mean weight[a] (kg)*	*Biomass (kg/km²)*
Marsupials			
Didelphis marsupialis[b,c]	55	0.83(16)	45
Philander andersoni [b,c]	25	0.27(9)	7
Metachirus nudicaudata[b,c]	12	0.28(12)	3
Caluromysiops irrupta[b]	10	0.25	3
Marmosa cinerea[b,c]	10	0.15(2)	1
Marmosa noctivaga[b,c]	15	0.06(2)	1
Bats			
Many genera and species	?	?	(50–100?)
Primates[d]			
Ateles paniscus	25	7.0	175
Alouatta seniculus	30	6.0	180
Cebus apella	40	2.6	104
Cebus albifrons	35	2.4	84
Pithecia monachus	2?	1.0	2
Saimiri sciureus	60	0.8	48
Aotus trivirgatus[e]	40	0.7	28
Callicebus moloch	24	0.7	17
Saguinus imperator	12	0.4	5
Saguinus fuscicollis	16	0.3	5
Cebuella pygmaea	5	0.1	1
Procyonids			
Nasua nasua	4	2.5	10
Potos flavus and[b,f] *Bassaricyon alleni*	50	2.0	100
Rodents			
Proechimys spp.[b,c]	270	0.23(11)	62
Oryzomys, etc. spp.[b,c]	550	0.064(34)	35
Dasyprocta variegata	8	2.0	16
Myoprocta pratii	4	0.5	2
Agouti paca[b,f]	24	6.2	150
Sciurus spp.[b]	25	0.6	15
Peccaries			
Tyassu pecari[g]	3	35	105
Tyassu tajacu[g]	5	25	125

(Table 3.1 cont.)

Taxon	*Approx. no. individuals per km²*	*Mean weight*[a] *(kg)*	*Biomass (kg/km²)*
Birds[h]			
Tinamous	120	0.5	60
Curassows, guans	30	1.5	45
Trumpeters	10	1.0	10
Pigeons	80	0.2	16
Parrots	200	0.1	20
Toucans	100	0.25	25
Passerines	?	<0.05	±25
Summary			
Marsupials	130		60
Bats	?		±75
Primates	275		650
Procyonids	80		110
Rodents	875		280
Peccaries	8		230
Birds	>570		±200
Total frugivore biomass			±1600

NOTE: Includes granivores and omnivores.

[a] Estimated mean weight over all age and sex classes, or adult weight when only adults were included in the census.

[b] Estimates taken from unpublished census results of Louise Emmons.

[c] Mean weights of animals of all ages trapped in the study area by L. Emmons with number of individuals weighed in parentheses.

[d] Primate densities were estimated by various methods in the census conducted in 1974. A full account of the methods and results is in Janson (1975).

[e] The density of *Aotus* is based on a complete count of groups made by Patricia Wright.

[f] Mean weights computed from data accompanying specimens in the U.S. National Museum.

[g] Density estimates from Kiltie and Terborgh (1983).

[h] Avian densities based on sketchy census data and educated guesswork.

terrestrial mammals (agoutis, acouchis, coatis, squirrels) and bats. Emmons has provided estimates of the former, but bats remain an unknown component of the whole picture. A rough indication of their potential role is given by Morrison (1978) who estimates the population density of *Artibeus jamaicensis* to be 6 ± 2 per ha on Barro Colorado Island, Panama. In many Neotropical forest localities *A. jamaicensis* is the most abundant fruit bat (at least in mist nets), often equaling the combined densities of all other frugivorous species. At a body weight of ca. 50 g, the biomass of this species on Barro Colorado Island is roughly 30 kg/km². If we assume very crudely that *A. jamaicensis* contributes half the biomass of chiropteran frugivores, then the total would fall somewhere in the range of 50–100 kg/km^2. Compared to the biomass of other frugivore groups at Cocha Cashu, this is a relatively low value, about equivalent to that contributed by marsupials, and less than 10% of the total.

Among the frugivores at Cocha Cashu, primates are unequivocally the leading group, contributing about 40% of the aggregate biomass. Other groups, in order of their relative importance are rodents (17%), peccaries (14%), birds (13%), procyonids (7%), bats (5%), and marsupials (4%). Within the more limited assemblage of arboreal frugivores (primates, birds, procyonids, and bats), primates are clearly preeminent (60%). Terrestrial frugivores tend to concentrate on nuts (peccaries, squirrels) and seeds (other rodents, tinamous), and in any case have available only those resources that arboreal species fail to harvest. Thus, the major potential competitors of primates are certainly other arboreal forms, principally birds and procyonids. Within this larger framework, the five species included in our study play a minor, though still appreciable, role (38% of the primate biomass; 23% of the arboreal frugivore biomass).

The data in Table 3.1 deserve further comment because of the paucity of information available on the structure of tropical forest mammal communities. Only two comparable surveys have been reported in the literature, one for a seasonal secondary forest in northern Venezuela (Guatopo: Eisenberg et al. 1979) and one for a mixed primary and secondary forest in Panama (Barro Colorado Island: Eisenberg and Thorington 1973). Estimated mammalian frugivore biomasses for these two localities and Cocha Cashu are given in Table 3.2. Each of the sites possesses some distinctive characteristics that appear to introduce distortions in the cross comparisons. For example, the Guatopo forest is semideciduous and is subject to hunting pressure. Perhaps because of these factors, the biomasses of primates and peccaries are very much lower than in the other two communities. By default, rodents contribute a disproportionate fraction (56%) of the total. The comparison between Barro Colorado Island and Cocha Cashu is inherently better because of great similarities in latitude, climate, generic composition of the forests and total annual fruitfall from the canopy (see Chapter 2). But Barro Colorado lacks a normal complement of top predators (no puma, jaguar, or harpy eagle), a deficiency that may have resulted in some major disruptions in the fauna (Terborgh 1980). Nevertheless, the total frugivore biomass does not appear to differ significantly from that at Cocha Cashu, though the balance of species does. There are many more howler monkeys, agoutis, and coatis at Barro Colorado, but fewer capuchins and spider monkeys. Squirrel monkeys are not present in that part of Panama, white-lipped peccaries disappeared from the island several decades ago, spider monkeys have only recently been reintroduced and have not yet reached carrying capacity, and kinkajous and olingos were

Table 3.2
Comparison of mammalian frugivore biomass at three Neotropical forest localities (kg/km²).

	Locality		
Group	*Guatopo, Venezuela*[a]	*Barro Colorado Is., Panama*[b]	*Cocha Cashu, Peru*
Marsupials	62	110	60[e]
Primates	167	421	650
Procyonids	18	120	110[e]
Rodents: large[c]	270	454	168
small	84	186	114[e]
Peccaries	34	153	230
Total	635	1444	1332
% Arboreal[d]	29	37	57

[a] Data from Eisenberg et al. 1979.
[b] Data from Eisenberg and Thorington 1973.
[c] Large rodents are agoutis, acouchis, and pacas; all others are included in the small category (rats, mice, squirrels).
[d] Here taken as primates plus procyonids.
[e] Some species in group not censused.

not censused. The most outstanding differences are a greater overall abundance of monkeys at Cocha Cashu and of large caviomorph rodents at Barro Colorado. To what extent these contrasts are due to intrinsic differences in the sites or to the lack of large predators at Barro Colorado cannot be ascertained at present.

Socioecological Characterizations of Cocha Cashu Primates

The forest at Cocha Cashu supports 11 species of primates. By extending the area of interest a few km beyond the limits of the trail system, two more species are included, bringing the total to 13 sympatric species. This is the richest primate community yet reported for a single New World locality, and rivals the diversity found in the equatorial forests of West Africa (Gautier-Hion et al. 1980; see also Chapter 10).

Since many readers will not be familiar with all the species, we introduce them on the following pages in a series of thumbnail sketches. Salient characteristics are summarized in Table 3.3. We shall begin with the 8 species not included in the study, and conclude with 5 that were.

Species Not Included in the Study

Ateles paniscus–Black Spider Monkey. This is the largest primate at Cocha Cashu, the males weighing up to 10 kg. Females are somewhat smaller, 6–8 kg, the dimorphism being fairly pronounced. *Ateles* are almost wholly frugivorous (Klein and Klein 1977), although some leaves are eaten in a rather casual fashion. We have not seen them hunt or consume animal prey. They live high in the forest, being particularly fond of the large spreading crowns that constitute the upper canopy. They move with acrobatic speed and agility, preferentially walking along horizontal limbs, but frequently switching to brachiation, especially when leaping or swinging from one crown to the next. When startled they can easily outdistance an observer on the ground.

Ateles have a complex and interesting social system characterized by a great fluidity and ephemerality of groupings. The only enduring social units are composed of females and their dependent young. As has been observed in other *Ateles* populations (Klein and Klein 1975), small pods of females, subadults, and juveniles commonly form short-term associations lasting from a few minutes to hours or even days. Large aggregations of twenty or more often form in the vicinity of major fruit concentrations. These usually include all age and sex classes. Adult males are often solitary or in small groups of varied composition. Male-female pairs are seen infrequently, a finding corroborated by the Kleins.

The existence of larger stable groupings analogous to the troops of other species is still somewhat problematical. In Colombia, Klein and Klein (1975) noted that recognizable individuals associated together in many combinations, but that the combinations formed closed sets. Members of different closed sets (= clans) were not seen intermingling. Our own observations of *Ateles* are not detailed enough to confirm or dispute the Kleins' findings, although we have witnessed seemingly aggressive shouting matches between opposing mixed-sex groups of *Ateles*. These matches ended with the separation of the groups in a manner that could be interpreted as mutual avoidance. Thus it seems probable that *Ateles*, like chimpanzees (Wrangham 1977), recognize larger groupings among themselves, even though they spend most of their time dispersed in much smaller units.

Ateles are opportunistic foragers, aggregating in sectors of the forest that offer high fruit densities, and later shifting to other such areas or dispersing widely to exploit individual trees. A variety of loud calls, some audible over hundreds of meters, may serve to signal the locations of fruit trees to widely scattered individuals. *Ateles* are particularly partial to figs, and are perennially

Table 3.3
Some characteristics of the primates found at or near Cocha Cashu.

Species	Approx. adult weight (kg)	Diet[a]	Position in forest[b]	Loco-motion[c]	Prehensile tail	Habitat specialist	Mean troop size	Social system[d]
Ateles paniscus	8.0	F,L,U,Ne	C	R,B	+	no	variable	FF
Lagothrix lagotricha	8.0	F,L?	C	R	+	no?	15	FF
Alouatta seniculus	8.0	F,L,U	C	R	+	no	6	PS
Cebus apella	3.0	F,I,Nu,Ne,P,U	M,C,G	R	+	no	10	PM
Cebus albifrons	2.8	F,I,Nu,Ne	M,C,G	R	(+)	no	15	PM
Pithecia monachus	1.5	F,?	?	R	–	yes	5?	?
Saimiri sciureus	0.9	F,I,Ne,Nu	M,U,C,G	R	–	no	35	PM
Aotus trivirgatus	0.8	F,I,Ne	M,?	R	–	no	4	M
Callicebus moloch	0.8	F,L,U	M	R	–	yes	3	M
Callimico goeldii	0.5	F,I?	U,?	V,R?	–	yes	3?	M?
Saguinus imperator	0.5	F,I,Ne	U,M,C	R,V	–	yes	4	M
Saguinus fuscicollis	0.4	F,I,Ne,S	U,M,C	V,R	–	yes	5	M
Cebuella pygmaea	0.1	S,I,Ne	U,M	V,R	–	yes	5	M

[a] F = ripe fruit; I = insects, etc.; L = leaves; Ne = nectar; Nu = nuts; P = pith; S = sap; U = unripe fruit.
[b] C = canopy; G = ground; M = midstory; U = understory.
[c] B = brachiate; R = branch run; V = vertical cling and leap.
[d] M = monogamous; PM = polygynous, multimale troops; PS = polygynous, single male troops; FF = fission-fusion.

found in greatest numbers in certain middle-to-late successional stands where figs are especially abundant. Dietary overlap with other large primates is considerable (see Table 5.3), though *Ateles* consume some species (e.g., several *Lauraceae*) that are avoided by most of the rest. Habitat selectivity is minimal, presumably due to the extraordinary mobility of *Ateles*, its broad diet, and opportunistic use of patchy resources. They use virtually every type of vegetation at some time in the year, wandering even into the early successional canebrakes along the riverbank to feed on *Cecropia* fruits (February–March) or *Erythrina* flowers (July–August).

The reproductive rate is low. Females are rarely accompanied by more than one dependent offspring. This implies an interbirth interval of at least two years. Our observations are too scanty to indicate a seasonal pattern of births.

Lagothrix lagotricha–Woolly Monkey. While inexplicably absent in the vicinity of Cocha Cashu, these large (8–10 kg) handsome monkeys are common on the right bank of the Manu River at Pakitza, some 20 km downstream. Local Machiguenga Indians report that they are also common far upstream. The reason for their patchy occurrence is unknown. Observations by Ramirez (1980) on the population at Pakitza indicate that *Lagothrix* feeds in large canopy trees on a variety of soft fruits, including figs and *Brosimum* (Moraceae). Some leaves are also included in the diet.

Woolly monkeys live in sizable groups of 15–20 individuals which may split into smaller subgroups when feeding. The home range is large, probably several km^2, though this is merely inferred from a limited number of observations. To date there has been no concerted field study of this interesting species, and little is known of its ecology.

Alouatta seniculus–Red Howler Monkey. Howlers, along with the previous two species, complete a trio of large 6–10 kg monkeys that frequent the high canopy of the forest. In habits, howlers could hardly differ more from the other two. Sedentary and inactive, they live in stable groups that inhabit small, indistinctly bounded home ranges. Though the troops contain from 3 to 7 individuals, the composition is consistent, including a single adult male, one or two reproductive females, and their accumulated immature offspring. Adjacent groups scrupulously avoid one another, and keep track of each other's respective positions by ritualized howling at dawn. When groups do meet, a prolonged howling match usually ensues, after which the participants tend to move in opposite directions.

Being partly folivorous, howlers live in slow motion. Filling their stomachs three or four times a day in relatively brief feeding bouts, they expend most of the remainder of their lives in shameless lethargy, sprawled out on a broad horizontal limb or curled in a comfortable fork (*A. palliata* spends 79% of the day at rest: Smith 1977). They move less (300–500 m/day) than any other primate in the community, except perhaps *Cebuella*. Their diet consists of young (flush) leaves and fruit in proportions that vary seasonally (Glander 1975; Milton 1979). No animal food is taken, nor do they visit flowers or resort to palm nuts as several other species do when soft fruit is scarce. More than any other primate in the community, howlers feed on immature fruits. It is probably this ability that tides them over the period of fruit scarcity, though quantitative data on this point are lacking. Like *Ateles*, howlers occupy all habitats at Cocha Cashu except the outermost canebrakes on the riverfront. They are the only monkeys that regularly descend to drink. In the dry season their need for water becomes so urgent that they will cross 100 m or more of open beach to reach the river, thereby exposing themselves to both aerial and terrestrial predators. Howlers are also the only monkeys to come to salt licks. We have seen them on several occasions eating exposed dirt in hollows at the bases of overturned trunks, and in peccary wallows.

Females appear to produce young at regular, circa yearly, intervals. We have not observed dispersal or the formation of new troops.

Callicebus moloch–Dusky Titi Monkey. This is a small brown monkey, about the size of a *Saimiri* (ca. 800 g), but extremely furry so that it appears larger. It is entirely vegetarian in diet, but possesses habits that diverge markedly from those of the species mentioned above.

Titis live in monogamous family units, usually composed of a male, female, and one or two juveniles. Even when tame, titis are exasperatingly difficult to follow because of their extreme crypticity. Not only do their somber colors blend into the shadowy background of the forest, but they have the disconcerting custom of sitting motionless in one spot for many minutes, and then moving silently to another location a few tree crowns away. Unlike the larger species described above, they climb into the high canopy only to feed, otherwise spending their time much lower in the thick and often viny midstory of the forest. Many of their food plants are small trees or vines which offer only part of a meal at a stop. Thus titis tend to move more and to use more trees per day than howlers, though between meals they show the same predisposition to lethargy.

The diet consists of small, fleshy fruits supplemented with leaves, especially vine and bamboo leaves (Kinzey and Gentry 1979). Unripe fruits are eaten occasionally, as are certain flowers (esp. *Bignoniaceae*). The preferential use of vine leaves may be particularly significant, because vines tend to grow continuously, rather than in periodic flushes as do most tropical trees. Young vine leaves thus constitute a continually renewing resource, while the expanding tree leaves consumed by howlers more closely approximate a nonrenewable resource. It is possible that this distinction is crucial to understanding the contrasting territorial systems of the two species (as will be discussed in more detail in Chapter 10).

Callicebus families vocally advertise and defend a definite area for their own exclusive use (Mason 1968). Each family announces its presence nearly every morning between the hours of six and eight in an elaborate antiphonal duet in which the male and female roles are precisely coordinated. Often the two animals sit side by side on a limb when performing, oriented in the direction of a countersinging pair of neighbors. Longstanding acquaintance with several families, and a limited amount of following of one (see Chapter 5) indicate that territorial overlap is minimal, corroborating Mason's (1968) findings with the same species in Colombia. Our observations differ from Mason's, however, on the matter of territory size. At Cocha Cashu, *C. moloch* territories fall in a range of 6–12 ha, more than ten times larger than Mason's reported mean of 0.5 ha. The *Callicebus* population Mason studied was in an isolated 7 ha forest patch where unusual densities could have resulted from reduced interspecific competition for food resources and/or reduced predation. Having casually observed *Callicebus moloch* in various parts of Peru, it is my impression that the larger territories of the Cocha Cashu population are more typical.

Callicebus families are unevenly distributed over the study area. They occupy a wide range of habitats but are absent from uniform expanses of high canopy forest. Their territories may contain a considerable proportion of such forest but invariably include other habitats as well. What all *C. moloch* territories seem to hold in common is a substantial amount of "edge." *Callicebus* families occupy nearly the entire margin of the lake, the riverfront, and the borders of all the swamps in the area. A few territories are scattered in the interior of the high forest but always in places where numerous treefalls have created a patchwork of openings and tangled successional vegetation. Such sites offer not only the high density of vines and thick cover that *C. moloch* seems to prefer, but probably an above-average concentration of food plants as well.

Pithecia monachus–Monk Saki. We know nothing of the habits of this species. A single group has been sighted three times in tall forest near the northern boundary of the study area. Its absence from the remainder of the area enclosed by the trail system suggests that it is highly selective of habitat, but the details of its requirements are unknown.

Aotus trivirgatus–Night Monkey. These colorful, wholly nocturnal monkeys occupy most, if not all, of the study area, as judged from the locations of calls and encounters when walking the trails at night. We find them in small, presumably family groups of three to five individuals. They move rather slowly through the forest, usually at midlevel, emitting frequent low sounds that must serve to keep the groups intact. Their diet includes many species of soft fruits that are consumed by diurnal primates. One distinctive habit is that of remaining for long periods in a single tree, resting in its branches between widely spaced feeding bouts (Janson et al. 1981). In a recent study elsewhere in Peru, it was found that *Aotus* groups show strong mutual repulsion and avoidance (Wright 1978). In their use of concentrated fruit resources, night monkeys resemble the larger monkey species more than *Callicebus*, to which they are closer in size and taxonomic position.

Callimico goeldii–Goeldi's Marmoset. We have encountered this reputedly rare species only twice, both times in swamp forest several km outside the limits of the regular study area. The groups consisted of 2–3 individuals that moved quickly through the dense understory by vertical clinging and leaping. We know nothing of the ecology of *Callimico* other than that it shares its restricted habitat with the two tamarins, *Saguinus fuscicollis* and *S. imperator*.

Cebuella pygmaea–Pygmy Marmoset. These remarkable gnome-like creatures are the smallest of all monkeys (ca. 100 g). Not previously known in southern Peru, they appear to be quite uncommon in the Manu region as, to date, we have discovered the locations of only three groups. Like other marmosets, they live in family units in which the males play a major role in caring for the young.

We casually observed one group over a period of 2½ years. During this period its numbers increased from 4 to 9 and decreased again subsequently. Twins were born in December 1976 and again in June 1977, presumably to the same female. The young were carried by the adult male, as is the rule among marmosets. In accord with previous reports (Castro and Soini 1977; Ramirez et al. 1977), the family occupied a tiny territory of less than 0.1 ha, and within this spent most of its time in just four or five trees. After the first year the group moved to a second location about 200 m from the first and remained there another year before moving again. Both sites were located on the bank of the Manu River. All of the *Cebuella* territories we have seen are centered around a small collection of trees that are exploited for sap. The adults chew characteristic holes in the trunk and major limbs of the favored trees, and visit them repeatedly in the course of a day. New holes are excavated on a daily basis, as the pits cease to yield after a few days. Thus, between January and June of 1977, the estimated number of holes in one tree increased from 300 to over 2000. Although such intensive and prolonged exploitation must impose a severe burden on the trees, most heal their wounds and survive for years after the *Cebuella* have abandoned them.

It appears that many species of trees yield sap of suitable quality, though only a limited selection is used in any single locality. At Cocha Cashu the favored species is *Inga mathewsiana* (Leguminosae), an abundant species along the lake and river margins. We have also found the telltale pits on species of *Coccoloba* (Polygonaceae), *Sloanea* (Elaeocarpaceae), and *Guarea* (Meliaceae). Where *Cebuella* has been studied in northern Peru, an entirely different collection of trees is exploited. These include species of *Quararibea* (Bombacaceae) and *Voichysea* (Voichyseaceae) (Ramirez et al. 1977).

If the nutritional requirements of a *Cebuella* family can be satisfied by the resources available within 0.1 ha, one wonders why the groups are so widely scattered. Certainly it is not due to a scarcity of sap trees. *Inga mathewsiana* is an abundant tree at Cocha Cashu, and is distributed in several habitats. A virtually continuous row of them overhangs the western shore of the lake, for example. Yet we have examined these and many other individual trees around the study area without finding any sign of *Cebuella*. Clearly, more is required than appropriate sap trees. One further criterion for suitable habitat may be adequate cover. Although some exploited sap trees are quite exposed, they are always close to dense viny tangles where the animals seek shelter when not feeding. The habitual nighttime roost site is especially well protected by vines. Moreover, there are two considerations which suggest that a *Cebuella* group needs a good deal more space than it actually occupies at any one time. One is the eventual need to move to a new site after one set of sap trees has been exploited to exhaustion. The other is a need for an alternative resource base during portions of the dry season. Prolonged droughts of several weeks' duration are a regular feature of the annual cycle in southeastern Peru. Sap flow into freshly made *Cebuella* pits and into simulated pits cut with a pocket knife is extremely meager during these dry spells. The group that we had under intermittent observation reduced the use of its sap trees during the dry season months of July and August and in substitution fed on the nectar of *Combretum assimile* vines (Janson et al. 1981). To do this, the animals were seen to move as much as 200 m beyond the limits of their normal territory. Thus, it is our impression that the space requirement of a *Cebuella* group includes dense vine thickets for shelter from predators, a dry season nectar source, and an adequate number of present as well as future sap trees. It may take several hundred meters of riverfront to meet these requirements, even though the animals may spend weeks or months on end living exclusively within a tiny core area around their current sap trees. The greater densities of *Cebuella* populations reported from northern Peru can be explained plausibly by a reliable year-round sap flow in the more equitable equatorial climate.

Species Included in the Study

I now introduce the 5 species of primates included in the study. Although they differ conspicuously among themselves in such important traits as size, group structure, and use of space, they form a coherent subset of the primate community on the criterion of diet. All are omnivores and, at a superficial level, their diets are very similar. Fleshy fruits constitute the bulk of their carbohydrate and caloric intake, while arthropods and other animal prey provide most of their protein requirements. The accounts that follow are brief because the remainder of the book is devoted to providing further details.

Cebus apella–Brown Capuchin. Largest of the five species, *Cebus apella* is a conspicuously burly animal which conveys a great sense of diligence and purpose as it forages. Adult males weigh well over 3 kg, and females somewhat less. The modal group size is 8–12 (rarely to 14) individuals; the typical composition includes an alpha and one or more subordinate adult males, one to four parous females, and assorted subadults, juveniles, and infants. Small subgroups of one to three subadult and/or low-ranking males drift from troop to troop and are occasionally encountered alone in the forest. At least 10 *C. apella* troops live partly or wholly within the study area.

Cebus albifrons–White-fronted Capuchin. Although slightly smaller than the preceding species, and noticeably more slender in build, *C. albifrons* regularly dominates its larger congener in confrontations over food. Interactions between the species are normally peaceful, however, and they may sometimes be seen feeding and traveling together in mixed associations. Troop structure is essentially the same as in *C. apella*, although the maximum group size is distinctly larger (up to 20; modal size around 14–16). In our experience, minimum troop size is about the same in the two species (ca. 8). Young males have been seen to switch troops, but have not been observed to form isolated subgroups as in *C. apella*. Three troops occupy the study area and a fourth makes occasional forays into it from the west. Although home ranges overlap broadly, troops show strong mutual avoidance mediated by loud vocalizations audible for at least 250 m. The two species of *Cebus* utilize all habitats present in the study area, though on differing temporal schedules.

Saimiri sciureus–Squirrel Monkey. This is the most abundant primate at Cocha Cashu, and the one with the largest troop size, typically between 30 and 40 individuals. These noisy throngs normally contain a balanced mix of males and females, adults and juveniles, though following the birth season (October–November) there may be a tendency for nursery groups of females and infants to separate from groups consisting predominately of males and subadults (cf. Thorington 1967).

Squirrel monkeys are exceedingly active, often foraging nonstop from dawn to dark. Smaller than *Cebus* (males ca. 1 kg, females ca. 800 g), they are able to traverse slender branches and vines as they filter through the dense midstory of the forest in search of insects. Although *Saimiri* habitually form mixed troops with either *Cebus* species, their home ranges are much more extensive than those of their larger associates. On many occasions we followed troops from one extremity of the study area to another, where, for lack of trails and familiarity with the terrain, we stayed behind while they continued for unknown distances beyond. *Saimiri* troops show no evident manifestations of territoriality, and it is commonplace to see two troops intermingling in the crown of a large fruit tree. Up to four troops may congregate in the vicinity of a concentrated supply of fruit, indicating broad overlap of home ranges. Because we have not succeeded in associating known troops with definite areas, the number of troops using the study area has not been determined, though it is at least four, and perhaps as many as eight.

Saguinus imperator–Emperor Tamarin. With its black bandit mask, pink nose, absurd mandarin mustaches, and orange tail, the emperor tamarin has the most arresting appearance of any New World monkey. Together with its less ornate congener, *S. fuscicollis*, this squirrel-sized (400–550 g) member of the marmoset family possesses a number of distinctive behavioral and ecological traits. It lives in small (2 to 10), extended-family groups, containing a single reproductive female and one or more adult males. Four, possibly five, strongly territorial groups inhabit the study area, but at a biomass density of less than a tenth of that of *Saimiri* or either *Cebus* species. Like *Callicebus*, with which they occasionally associate, both *Saguinus* species advertise their presence by means of loud vocalizations, especially near territorial boundaries. These challenges frequently provoke responses in kind from territorial neighbors, and close range confrontations, which sometimes escalate to physical combat, regularly ensue. At most times, however, the tamarins lead extremely cryptic lives, spending several hours each day in the security of thick cover.

Saguinus fuscicollis–Saddle-backed Tamarin. Slightly smaller and less dramatic in appearance than its congener, this species displays virtually identical social and territorial habits. A most surprising discovery was that groups of the two species occupy commonly held territories. Within the jointly defended boundaries, the groups live in loose association, often but not always foraging together, and maintaining contact through vocalizations. Both species participate in calling at territorial boundaries; in any confrontation that may follow, each directs its aggression toward conspecifics in the opposing mixed group. This extraordinary behavior will be discussed in greater depth in Chapter 8.

Summary

At this point it may be helpful to summarize the salient characteristics of the species we studied with respect to body size, ranging behavior, territoriality, and group size. At the large end of the size scale are the two capuchins (*Cebus* spp.), each weighing about 3 kg as adults. The squirrel monkey (*Saimiri*) falls in the middle at 800–1000 g, while the tamarins (*Saguinus* spp.) lie at the low end of the scale at 400–550 g adult weight. *Saimiri* have almost indefinitely large home ranges that are not advertised or defended in any discernible way, as two groups often feed together in large fruit trees without any show of antagonism. The *Cebus* spp. have smaller and more discrete home ranges, though at 150 ha (*albifrons*) and 70 ha (*apella*) their home ranges are still much larger than those of many Old World monkeys of even greater body size. There is strong mutual avoidance by *albifrons* groups, but only mild avoidance or apparent indifference when *apella* groups meet. The two tamarins, which live in mutualistic mixed associations, live within smaller nonoverlapping territories that are vigorously advertised and defended. Group sizes of the five species follow the same order as their space requirements, with *Saimiri* having the largest groups (30–40), followed by *Cebus albifrons* (12–20), *C. apella* (8–12), and the *Saguinus* spp. (2–10).

It is our working hypothesis that many of these differences can be understood as adaptations for exploiting food resources with distinct spatial and temporal patterns of abundance. To examine the hypothesis we need to know both about the availability of resources to the animals, and about how the animals exploit them. The fruit-trap data presented in the preceding chapter gave a broad overview of the seasonal pattern of fruit abundance at the study site, but they imply nothing about how frugivores may cope with annual periods of scarcity, or how resources may be exploited in different ways by potentially competing species. These topics constitute recurrent themes that will be discussed in a variety of contexts throughout the book.

4 Activity Patterns

The nature of an animal's diet imposes many constraints on its use of time and space. In this chapter I shall consider the use of time by the five species, particularly as it relates to their food-finding needs. Being omnivores, they stand between herbivores and insectivorous predators in trophic position, and in this we should expect to see that their behavior is intermediate, and perhaps in some ways even compromised by the opposing exigencies of the two ways of life.

Among herbivores, the amount of time an animal devotes to feeding is conspicuously related to body size (Clutton-Brock and Harvey 1977a). Large ungulates, for example, commonly spend more than half of their time browsing or grazing, and are often active by night as well as by day (Gautier-Hion et al. 1980; Emmons et al. 1983). This is true of the largest mammals of the Neotropical forest: tapirs (Terwilliger 1978), deer, peccaries (Kiltie 1980), and their principal predators, jaguars and pumas. Small herbivores such as voles, rabbits, and folivorous primates forage intermittently and are generally either strongly diurnal or nocturnal. This is because they are able to ingest food much more rapidly than their guts can process it. Rest and various forms of social activity may occupy half or more of their daily time budgets (Hladik and Charles-Dominique 1974; Oates 1977; Smith 1977; Sussman 1977). With large species, ingestion is slower relative to metabolic needs, and food gathering activities predominate.

Quite a different set of constraints operate on insectivores. Food items are small, well scattered in the environment, and often camouflaged or concealed. Rates of ingestion do not greatly exceed rates of digestion, and searching is the primary time-consuming activity. While it is clear that a bison can graze at a much higher absolute rate than a vole, there is no corresponding scalar relationship between body size and prey capture rates among insectivores. To the contrary, among birds at least, capture rates decline markedly with body size, although, to be sure, large birds take larger prey than small birds (Fitzpatrick 1978). A definite upper limit to the size of insects, and an increasing scarcity of numbers in successively larger size classes, act together in imposing an upper limit to the size of obligate insectivores (Schoener and Janzen 1968; Janzen 1973a). Among birds this limit is in the range of 100–200 g (Terborgh 1980), and practically all of the larger species (<100 g) are nonpasserines, having reduced metabolic rates relative to members of the passerine order that includes most small avian insectivores. Very much the same story pertains to mammals. Nearly all obligately insectivorous bats, shrews, marsupials, tree shrews, tarsiers, etc. weigh less than 100 g (Emmons et al. 1983). Larger forms such as skunks, opossums, viverrids, procyonids, and galagos are either omnivorous or include vertebrate prey in their diets (Charles-Dominique 1974). The exceptions are a few highly specialized animals that feed on social insects (anteaters, aardvark). For these, effective prey size is correspondingly large, since the prey units consist of whole colonies of ants and termites.

In view of the constraints that time and rates of energy acquisition impose on the size of insectivores, one can expect that there will be no wholly insectivorous monkeys. Indeed, this appears to be true (Jolly 1972). The smallest monkey, *Cebuella pygmaea*, weighs about 100 g and is a specialized sap feeder. Nevertheless, the size range of monkeys (from 0.1 to ca. 40 kg) is such that many of the smaller ones are not much larger than the largest insectivores. This opens the possibility of omnivory, which is in fact the dietary mode of most of the small to medium-sized primates (Jolly 1972).

As a life style, omnivory entails compromises in both behavior and diet (McNab 1980). A need to obtain protein in the form of insects puts major demands on the time budget, while a need to obtain carbohydrate in the form of fruit largely dictates the use of space. The digestive system is adapted to handling a wide array of high quality nontoxic food items. Relatively small volumes of material are processed relatively quickly. This state contrasts sharply with that of most species which obtain their protein from vegetable matter. These must process large quantities of often toxic or poor quality material slowly to allow for detoxification and breakdown of polysaccharides (Milton 1981). Thus a true omnivore cannot facultatively switch to being a vegetarian, even though it does enjoy a certain amount of flexibility in being able to vary the proportions of insects and fruit in the diet in accordance with fluctuations in resource availability.

It is clear that omnivory imposes constraints on the use of time and space because insects are diffusely distributed in the environment, and are hard to find, while fruit occurs in concentrated but widely scattered patches and may or may not be hard to find. The present chapter shall be devoted to an analysis of the time budgets of the five species and Chapter 7 will consider their use of space, especially as related to harvesting fruit. After an exposition of the method employed, the text continues with a comparison of the time budgets of the five species *inter se*, and collectively with a vegetarian primate in the same body-size range, *Callicebus moloch*. Daily activity cycles are next examined for significant patterns, and finally, seasonal variations in time budgets are considered in relation to variations in day length and fruit abundance.

Methods

Schedule of Samples

It was mentioned in the first chapter that we employed long sample periods, commonly 15–20 days, in order to resolve the large-scale ranging patterns of our animals. This resulted in sampling each species about once every three months. To keep to this schedule, the two *Saguinus* species were sampled together, with 10 days being devoted to each one. Shorter samples were more appropriate in this case because *Saguinus* territories are relatively small, but particularly because the focal troops of each species jointly occupied the same territory throughout most of the study.

With four out of the five species, we selected a particular study troop, generally the one living closest to the station. Though in some cases the selection of troops was conditioned on their being tame enough to follow, the choices were fortunate in that all the focal troops broadly overlapped in their use of the central portion of the study area. Only with the squirrel monkey (*Saimiri*) could we not be certain that we were observing the same troop in successive samples. This was only partly due to a paucity of distinctive animals that could have served as markers for particular troops. The seminomadic habits of the species presented a far greater difficulty in that the groups persisted in traveling hundreds of meters beyond the outer limits of the trail system, whereupon we would be obliged to abandon them. When this happened, we simply located another troop nearer the center of the study area and continued with that one, usually the same day. The actual schedule of observations, along with ancillary information on the composition of the study troops, is given in Table 4.1.

Samples were conducted with the collaboration of two to five observers. Expediency required that the person who ''put the monkeys to bed'' one evening also started out with them the next morning. When two people alternated in shifts, they exchanged places every three hours, communicating by means of two-way citizens' band radios. When three or four people participated, we used somewhat longer four-hour shifts. And for a single month (June 1977), when there were five observers, we engaged in a complex system of rotations that allowed us to maintain continuous contact with two troops at once.

A comment is due at this point on the monkeys themselves. Most of the troops at Cocha Cashu were never particularly shy. With a few notable exceptions, we were able to follow and maintain contact with any desired group right from the start. The initial reaction in most cases was one of mild avoidance. Individuals were easily displaced from their activities if approached too closely, and tended to remain higher in the forest than they normally would. Gradually, as they became accustomed to our presence, they lost virtually all of their concern. By the latter half of the year, individuals from all our study troops would come within two or three meters of us as they walked along branches or vines, or would descend to the ground practically at our feet. At no time did we intentionally interfere with their activities or attempt to touch or feed them. We wished to be regarded as neutral hangers-on, and to all outward appearances we were.

Measurement of Activity Budgets

In recording the activities of animals such as monkeys that engage in complex social interactions and that exhibit a wide range of behaviors, one must decide at the outset what to emphasize and what to neglect. It is simply impossible to record everything, and even if it were possible, the torrent of notes that would result would lack the virtue of systematic organization. Given that we were committed to documenting the feeding, foraging, and ranging behaviors of the species as faithfully and as completely as possible, other aspects of the animals' lives, such as the nuances of their social behavior, had to take a back seat in our protocol. In order to emphasize the ecological aspects of behavior, we used just five very broad categories to characterize activity states: travel, rest, feed (here, for convenience, used only in reference to consumption of plant materials), forage (to comprehend all the activities involved in finding and ingesting animal matter), and miscellaneous (includes activities that do not fall conveniently into the other categories, such as allogrooming, play, sexual and territorial behavior, predator mobbing and avoidance, and intragroup aggression).

Selection of a suitable sampling method for quantifying these activity states was complicated by a number of limitations and requirements. Compromises were necessary because any scheme we chose had to be uniformly applied to five, not just one, species, and among the five there were great differences in the number of animals per troop, the typical degree of dispersion while traveling or foraging, and the visibility of troop members in the dense foliage of the forest. Focal animal methods were precluded by the fact that in three out of five species (*Saimiri* and the two *Saguinus*) we were unable to distinguish individuals with any reliability.

Scan sampling offered the only practical alternative in the face of these limitations. In its strictest form, scan sampling calls for recording the activity of each troop member at regular (e.g., 10 or 15-minute) intervals. This, too, posed difficulties. For one, it was seldom possible at any given time to see all the members of a troop in the thick cover of the forest. For another, recording activities at regular intervals would have conflicted seriously with our program for documenting foraging behavior. This called for recording

Table 4.1
Dates of samples and composition of study troops.

Species	Dates of samples	Troop composition								Remarks
		No. individs.	Ad. males	Ad. females	Subad. males	Subad. females	Juv. males	Juv. females	Infants	
Cebus apella	Aug. 27–Sept. 8, 1976	16	6	3	2	2		1	2	Several supernumerary males.
Cebus apella	Oct. 14–Nov. 3, 1976	16	6	3	2	2		1	2	Baby born late Nov.
Cebus apella	Jan. 26–Feb. 16, 1977	11	3	3	1	1	2		1	3 Ad. males, 2 Subad. females, 1 Subad. male missing.
Cebus apella	April 2–April 8, 1977	11	3	3	1	1	2		1	Infants become juv.
Cebus apella	June 12–July 1, 1977	12→11	3→2	4	1	1	3			New ad. female in troop. Alpha male killed 6/8.
Cebus albifrons	Aug. 4–Aug. 13, 1975	8	2	2	2	1	1			
Cebus albifrons	Nov. 21–Dec. 8, 1976	8	2	2	2	1	1			One female pregnant. Baby born late Dec.
Cebus albifrons	Mar. 16–April 5, 1977	9	2	2	2	1	1		1	
Cebus albifrons	June 16–July 1, 1977	9	2	2	2	1	1		1	
Saimiri sciureus	Sept. 19–Oct. 10, 1976	35 ± 5								Babies born in Oct.

(*Table 4.1 cont.*)

Species	*Dates of samples*	*Troop composition*								*Remarks*
		No. individs.	*Ad. males*	*Ad. females*	*Subad. males*	*Subad. females*	*Juv. males*	*Juv. females*	*Infants*	
Saimiri sciureus	Feb. 17–Mar. 12, 1977	35 ± 5								
Saimiri sciureus	May 9–May 29, 1977	35 ± 5								Intense sexual competition among males.
Saimiri sciureus	July 22–Aug. 8, 1977	35 ± 5								
Saguinus imperator	Aug. 14–Aug. 23, 1975	3	1	1	1					
Saguinus imperator	Oct. 30–Nov. 2, 1976	4	1	1	1	1				Female pregnant. Twins born late Nov.
Saguinus imperator	April 9–April 22, 1977	5	1	1	1	1	1			One of twins missing.
Saguinus imperator	July 14–July 22, 1977	4	1	1		1	1			Ad. and Subad. males missing. Replaced by new ad. male.
Saguinus fuscicollis	Nov. 9–Nov. 17, 1976	5	1	1	1				2	Recently born twins present.
Saguinus fuscicollis	April 22–May 1, 1977	4→3	1	1	1			1→0		One twin missing, the other killed by a hawk on 4/25.
Saguinus fuscicollis	July 5–July 14, 1977	3	1	1	1					
Callicebus moloch	May 4–May 9, 1977	3	1	1				1		

sequences of search and capture actions in timed intervals of indefinite length (see Chapter 6 for details). The observer began timing a sequence whenever a foraging animal came into good view. It was impossible to concentrate on the foraging of individual animals and keep track of the distribution of activity states in the whole troop at the same time. The two types of data had to be taken sequentially, but the indefinite lengths of the foraging sequences prevented the observer from turning his/her attention to activity states on a regular schedule. Thus we settled on a protocol in which feeding and foraging measurements had priority. Whenever the observer's attention was not occupied with these, an assessment of the ongoing activity in the troop could be made. In practice, this meant that such assessments were made very frequently, every few minutes. Changes in activity state were then recorded as they occurred. The activity was regarded as continuous until the next recorded change. By summing up the intervals between recorded changes in activity state, a complete activity budget was compiled.

Some practical details of the procedure warrant further clarification. Assessment of the distribution of activities within a troop could seldom be based on a complete survey of its members. Large troops (*Cebus* spp. and *Saimiri*) were often spread out over a diameter of 100 m or more, making it impossible to check on all the individuals. Moreover, some animals were invariably hidden by foliage. While following squirrel monkeys, for example, one usually could count on having 5 to 10 individuals in view, but this meant that the other three-fourths of the troop was out of sight.

The tamarins presented a different sort of difficulty in their cryptic habits. A major part of every day is spent in safe retreats where the animals are secure from attacks by aerial predators. These retreats are commonly a massive vine tangle enveloping the crown of a large tree. Even though tamarins are effectively invisible in such places, one can safely conclude that they are at rest by elimination of all other possibilities (i.e., they are manifestly not feeding, traveling, or insect foraging). During periods of activity, it was usually possible to see what all the members of a tamarin troop were doing, because there were only three to five animals to keep track of and they tended to stick closely together.

When there was more than one concurrent activity, each activity was assigned a fractional value according to the assessment of the observer. For instance insect foraging routinely occurred during travel between fruit trees. In such cases, assessment of the activity state depends on the rate of travel as well as on the apparent degree of dedication to foraging. During rapid travel there is no time to forage, so the interval is scored unambiguously as pure travel. Often there was intensive foraging while the troop as a whole remained stationary. This was easily scored as pure foraging. In between these two extremes there is a spectrum of possibilities. We found it practical to recognize only two intermediate conditions, what we termed forage-travel (scored as ⅔ forage, ⅓ travel) and travel-forage (scored as ⅔ travel, ⅓ forage). By this convention we recognized three intensity levels of foraging activity and three levels of travel. How the activity state was scored at any time was based on the observer's evaluation of the activities of the animals in view, something that necessarily entailed a practiced judgment.

Similarly, multiple activities occurred in other combinations, e.g., rest-forage, rest-feed, feed-forage. To assess the distribution of activities, the observer determined the activities of as many troop members as possible and then assigned fractional values. (We kept track of feeding activity in a more meticulous fashion, as described below.) When some of the members of a troop were resting, there was a greater chance than at other times of underestimating the activity because of the inherent inconspicuousness of resting animals. However, resting occurred only at certain times, i.e., when the troop was not traveling. When an observer noticed that some animals were resting, a special effort was made to locate as many troop members as possible and to determine their status. The resulting activity budgets are certainly not as precise as those done with focal animal studies of highly visible species, such as baboons, but we feel they are reasonably accurate given the handicap of working in dense forest. Most importantly, the method was uniformly applied to all five species, thereby assuring the comparability of the results *inter se*.

Accounting of the time spent feeding in fruit trees was done by a more exacting procedure. We made a concerted effort to keep track of when animals entered and exited a feeding tree, and how many were feeding at all times. (In the case of *Saimiri*, with 30 to 40 animals present, this was not practical, and the results are accordingly more approximate.) Normally, the entire period between entering and leaving a tree was allocated to feeding, except, rather infrequently, when some of the animals elected to remain in the crown for a rest session. Feeding time in each visit to a tree was summed according to the number of animals feeding over the entire period of the visit to give a total number of monkey-minutes of feeding. (E.g., 7:12–7:14, two individuals feeding; 7:15, three now feeding; 7:16–7:17, five feeding, etc., for a total of $3 \times 2 + 1 \times 3 + 2 \times 5 = 19$ monkey-minutes.) These totals were then divided by the number of individuals in the troop to convert the figures to troop-minutes. The number of troop-minutes accumulated over a day was the feeding time for that day.

Annual Time Budgets

Average activity budgets derived by combining all samples for each species are shown in Figure 4.1. In comparing the species, the most obvious pattern is the pronounced reciprocity between insect foraging and rest. Insectivory is highly developed in the five omnivores, apparently far more so than in any Old World monkey except the talapoin (*Miopithecus talapoin*: Gautier-Hion 1973). Three of the species (*Cebus apella*, *C. albifrons*, and *Saimiri*) spend nearly half their waking lives searching for arthropods and other small prey. The two tamarins forage less and rest more, but this may be more a consequence of their smaller size than of a diminished representation of animal food in their diets (see Chapter 6). Lastly, the vegetarian titi monkey (*Callicebus moloch*) scarcely insect forages at all, and, in keeping with other species of similar dietary predilections, spends most of the day at rest (cf. Clutton-Brock and Harvey 1977a).

Unfortunately, there is as yet no empirical formula for calibrating the amount of time spent on different kinds of feeding activity against the amount of food actually consumed (Hladik 1977). Time budget data are available for

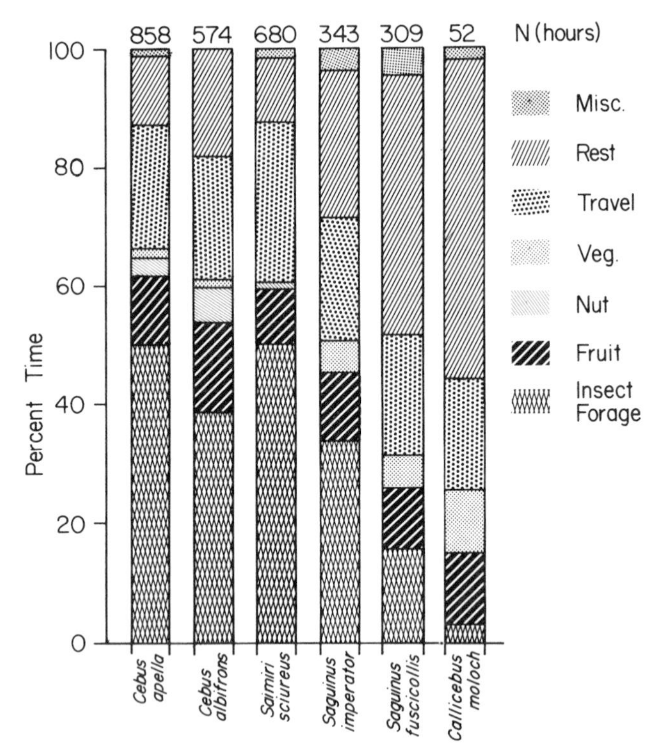

Figure 4.1 Activity budgets averaged over all samples. Numbers at top of columns are the total contact hours for each species. The miscellaneous category includes grooming, play, intraspecific aggression, sexual behavior, and territorial activity.

a substantial number of primate species, and stomach content analysis has been undertaken for a few, but there is almost no overlap in the two sets of species (Clutton-Brock and Harvey 1977a). Thus, it is known from the stomachs of collected specimens that about 50% of the intake of talapoins consists of arthropods, but it is not known how much time they devote to searching for them (Gautier-Hion 1973). Similarly, Charles-Dominique (1974) has shown insectivory to be highly developed in certain lorisids, but activity budgets are not readily assessed in these nocturnal animals.

It is nevertheless evident that capturing arthropods is a time-intensive occupation. Small insectivorous birds weighing between 8 and 49 g are compelled to spend 8 to 10 hours a day foraging in the forest at Cocha Cashu (Munn and Terborgh 1980). At times when there is little or no ripe fruit available, *Saimiri* forage virtually nonstop from dawn to dusk. It seems likely that talapoins, which are about a third larger than *Saimiri*, will prove to be equally intense in their foraging behavior.

Table 4.2 reveals another pattern that is common to all the species. It is that the sum of time spent traveling and time spent feeding on plant material is a nearly constant 40% for all six species (range 36–43%). Within the limits of variation, *Saimiri* spends the most time traveling (27%), and *Callicebus*, the vegetarian, spends the most time feeding (23%), though it spends no more time eating fruit (18%) than the rest of the species. Of course, travel and feeding are intimately coupled through the necessity of moving from one fruit

Table 4.2
Percent of daily time budgets devoted to various activities in six Neotropical primates.

Species	*Rest*	*Travel*	*Feeding on plant material*	*Travel and plants*	*Insect forage*	*Total feeding: plants and insects*
Cebus apella	12	21	16	37	50	66
Cebus albifrons	18	21	22	43	39	61
Saimiri sciureus	11	27	11	38	50	61
Saguinus imperator	25	21	17	38	34	51
Saguinus fuscicollis	44	20	16	36	16	32
Callicebus moloch	54	19	23	42	3	26

NOTE: Mutually exclusive columns do not sum to 100 because miscellaneous category was deleted.

tree to another. In addition, there are other considerations that bear on both of these activities.

The range of feeding time for all the species (11–23%) falls at the lower end of the times for other primates, as compiled by Clutton-Brock and Harvey (1977a). Several factors contribute to the lack of accord with other data. One is the small size of our species. Most of the other primates for which activity budgets are available are large Old World forms with predominantly vegetarian diets. All but one of our species are omnivores, and therefore consume plant material in smaller proportions than strict vegetarians. Finally, "feeding time," as represented in the data compiled by Clutton-Brock and Harvey, includes unknown, and certainly variable, amounts of time in the category we have called insect foraging. If we add this to the time spent feeding on plant materials, the total time for all food gathering activities jumps to very high values. Several of these values fall at or beyond the upper limits of reported feeding times for vegetarian species (right-hand column in Table 4.2).

To summarize the trends in these comparisons, small body size tends to facilitate increasing levels of insectivory by reducing the required overall food intake, and with it the amount of time spent feeding on plant parts. The additional uncommitted time can then be devoted to foraging, a task that offers low but relatively steady yields.

Generalizations about travel time are made less easily. Some primates return nightly (or daily, as with *Aotus*: Wright 1978) to a traditional roost tree, while most sleep wherever they find themselves at dusk. Requirements other than the need to travel between fruit trees may constrain or encourage movement. Baboons, for example, may have to move long distances to visit water holes (Altmann 1974). Among our set of species, *Saimiri*, and to a lesser degree *Cebus*, make extended forays for which the only discernible purpose is insect foraging. Harvest of both the animal and plant components of their diets thus seems to call for a substantial amount of travel. The tamarins, in contrast, appear to do most of their insect foraging while en route between fruit trees. But unlike the larger Cebids, tamarins are intensively territorial, and regularly patrol the outer bounds of their domains, a habit which adds appreciably to their daily excursions. Tamarins also possess preferred roost trees, from and to which they may journey 100 m or more at the beginning and end of a day. It is clear from these observations that troop movements, especially in tamarins, are conditioned by more than food-finding requirements. Moreover, the several species travel at characteristically different rates, so that the lengths of their daily excursions vary proportionately more than the amount of time spent in travel. *Cebus* and *Saimiri* tend to progress at a steady pace for long periods, while the tamarins and *Callicebus* typically move in a stop-go fashion that perhaps reflects a greater alertness to the threat of predators.

In spite of all these differences, the travel times of the six species are strikingly convergent. More than anything else, this must derive from their common need to visit a number of scattered fruit trees in the course of each day. Additional requirements, such as insect foraging in *Cebus* and *Saimiri* and territorial patrolling in the tamarins, apparently increase the travel budgets of all the species by similar amounts.

The category of behavior labeled "miscellaneous" in Figure 4.1 is, as mentioned previously, a catch-all for several types of social interactions: allogrooming, play, sexual behavior, and territorial advertisement and defense. The amounts of time involved are small and are represented only approximately because these activities seldom engage all the members of a troop at once.

For the *Cebus* species and *Saimiri*, the total amounts of time involved are minute (<2%). This is because none of them is either particularly social or territorial. Most of the total for *Cebus* is comprised of allogrooming bouts, the majority of which involve the dominant male or mother-offspring combinations.

Saimiri are the least social of all the species and rarely engage in allogrooming. Intimacy among them is confined to brief periods of huddling in small knots of two to several individuals (usually females and/or juveniles), or to roughhouse sessions among juveniles. Adult males, during the rather concentrated mating period, largely refrain from their usual foraging activities to engage each other in incessant dominance interactions and mutual sexual displays. But for most of the year, and for most of every day, *Saimiri* operate as individuals, each one selecting its own more or less independent pathway as it wends its way through the foliage in search of prey. A strong commitment to insectivory entails a nearly obsessive dedication; there is little leisure for social pursuits. This is nearly as true for *Cebus* as it is for *Saimiri*.

Callicebus and the tamarins represent quite a different situation. All of these live in closely integrated family units and rest a substantial portion of each day. Allogrooming is a frequent, indeed daily, pastime that to varying degrees engages every member of the group. Juvenile tamarins play among themselves on occasion, but for shorter periods and less frequently than young *Cebus*. We saw almost no play in *Callicebus*, but that was perhaps because the juvenile was not at a propitious age during the very brief sample (Patricia Wright, pers. com.).

Territoriality is highly developed in *Callicebus* and the tamarins, and plays a major role in their lives (Mason 1968; Neyman 1977; Dawson 1977). *Callicebus* pairs engage in loud and elaborate antiphonal duets nearly every morning between the hours of six and eight. The initiation of singing by one group triggers responses by neighboring groups within audible distance until most or all of the units in the local population have announced their whereabouts. These vocal challenges sometimes provoke adjacent families to rush together at territorial boundaries, but the contestants generally stop short when within sight of one another, and have not been seen to escalate the confrontation to the level of physical combat. Territorial maintenance in *Callicebus* is thus highly ritualized and perfunctory; the cost in time and energy is minimal.

Not so with the tamarins. Their territories are considerably larger than those of *Callicebus* (up to 70 ha; see Chapter 7), and consequently are more difficult to advertise and defend. While in the centers of their territories, neighboring groups may be as much as a kilometer apart, well beyond the effective limits of vocal communication. Advertisement is not spontaneous and ritualized, as it is in *Callicebus* and *Alouatta*, but appears to involve plan and purpose. Groups visit their boundaries on a regular, though infrequent, schedule. When a boundary is approached, loud calling begins. Individual calling bouts are brief, lasting only a few seconds, but are repeated at irregular intervals over periods of up to several hours. If no answer is obtained, the group will generally withdraw, either later the same day, or after another try on the following morning.

An answer from the group being challenged stimulates much intensified vocal activity. At times one or both groups will rush forward, and at others a prolonged vocal exchange will ensue. But whether the reaction is immediate or delayed, the level of excitement usually builds until the opposing groups make visual contact. Responses differ at this point. At one extreme the two sides may engage in a physical struggle. Small cliques of individuals rush excitedly back and forth over the imaginary dividing line, in full pursuit or retreat. Head-on clashes are often resolved by grappling and biting, and are sometimes so vigorously entered that the participants fall to the ground. At the other extreme, the confrontations lead to nothing more than extended vocal contests which wax and wane in intensity through several periods of climactic excitement before one or both sides show signs of diminished interest.

While territorial confrontations such as these are by no means daily events, when they do occur (ca. once a week) they may continue intermittently for as long as four to six hours. And even when a calling session arouses no response from its neighbors, a group seems to go out of its way and to devote a considerable amount of time (scored as travel and rest) to patrolling and advertising its borders. Thus, unlike the situation in *Callicebus*, territoriality in tamarins does entail major costs in time and energy—costs that are understated by our accounting procedure in which only the time spent in face-to-face confrontations is entered into the miscellaneous category.

Patterns in the Daily Activity Cycle

A pattern of activity common to all six species emerges when the seasonally averaged activity budgets are broken down by time of day (Fig. 4.2). Upon leaving the roost area in the early morning, the animals tend to move to a nearby fruit tree. One to several trees may be visited in fairly rapid succession, resulting in a peak of travel and fruit-feeding activity. Then, with the exception of *Callicebus*, which rests, the focal activity shifts to insect foraging. This continues, interspersed with periods of rest, throughout the middle part of the day. Finally, there is a more or less marked resumption of fruit-tree visitation in the late afternoon, accompanied by more concerted travel between stops. Similar activity cycles have been noted in a number of other primate studies (refs. in Clutton-Brock 1977 and see discussion in Raemaekers and Chivers 1980).

The pattern is easily rationalized. Fruit is consumed at the two times of day when calories are most needed: early in the morning when stomachs are

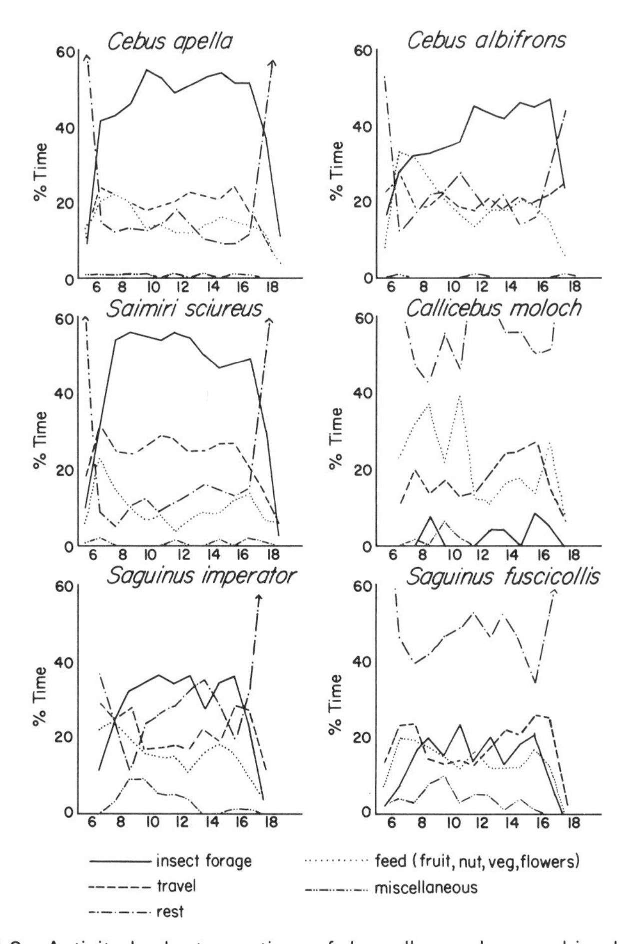

Figure 4.2 Activity budgets vs. time of day, all samples combined.

empty, and late in the day when a full stomach is needed to tide the animal through the long tropical night. Insects, which are a high value, low volume resource, are easily accommodated on top of a meal of fruit. Searching for them occupies the main part of the day. On this point our accounting system, in strictly separating foraging and travel, gives a somewhat distorted impression, because much of the travel that accompanies foraging is directly related to foraging. Thus, *Saimiri* and *Cebus apella*, which are shown to forage about 50% of the time and to travel about 25% through the middle hours of the day, might be represented more fairly as foraging nearly 75% of the time. But as it is operationally difficult to distinguish fruit-related from forage-related travel, we chose to classify all travel in a separate category.

Rest is a less ambiguous activity state in that it overlaps little with the other categories in the classification. Nevertheless, it does subsume such inactive behaviors as scanning and vigilance which, though ostensibly involve rest, are not necessarily restful. Rest tends to peak in the middle of the day, presumably because the animals are likely to be full by then, and because activity at high ambient temperatures may not be any more pleasant for monkeys than it is for men.

Seasonal Patterns of Activity

Cocha Cashu is far enough away from the equator so that there is an appreciable change in day length through the seasons (ca. 1.5 hr). The extra time this gives the animals between September and March could be used merely to expand the daily period of rest, or it could be invested in more serious pursuits such as providing for offspring. Although the evidence on this point is mainly circumstantial and anecdotal, it suggests that the animals do compromise by both resting and foraging a bit more than they do when days are short.

There are other mitigating circumstances than day length which may bear on the seasonal modification of time budgets. One is the timing of births. In *Saimiri* and the *Saguinus* species, most of the young arrive in a span of two months during the earliest part of the rainy season (October–November). A few births occur outside this period, but these are a definite minority. Thus, from September onward for several months, *Saimiri* and *Saguinus* females are burdened with the extra metabolic demands of pregnancy and lactation.

Another consideration is the food supply. Less travel between fruit trees may be required when fruit is relatively abundant, a time that also coincides with the advent of long days (cf. Chapter 2). This suggests that leisure may become a substitute for travel, and perhaps even replace some foraging when it is an easy matter to fill up on the most preferred of a wide selection of fruits.

Let us see what the monkeys actually do. The data are contained in Figure 4.3 but they give no clue as to the circumstances that pertain to the individual samples.

Cebus apella. Rest was minimal in the two dry-season samples, June and August, when the days were relatively short. Insect foraging was especially prominent in June, apparently in compensation for a paucity of fruit. February represents the opposite extreme: long days and maximal availability of fruit. The animals rested more and traveled less. Foraging continued at normal levels.

Cebus albifrons. The level of foraging activity remained virtually constant over all seasons, while adjustments were apparent in other categories. Again, rest was proportionately less in the two dry-season samples (June, August) than in the two taken during the wet season (November, March). The ex-

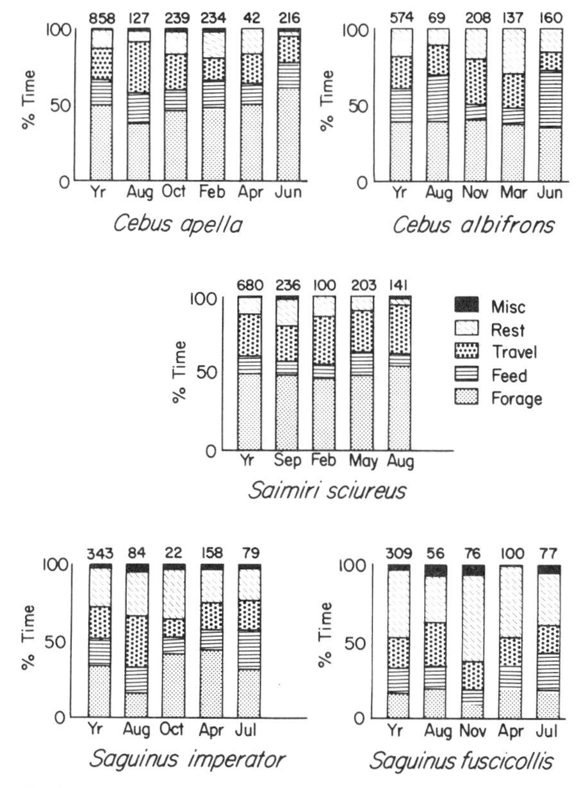

Figure 4.3 Activity budgets for each species by sample. Numbers at top of columns give the contact hours for each sample.

ceptional amount of time given to rest in March (29%) is explained by the many prolonged rains that occurred during the sample. In bad weather the monkeys tend to remain huddled in the shelter of an overhanging branch or vine tangle. Feeding time increased dramatically in June, paradoxically when fruit availability was near its annual nadir. This was because their main alternative food resource, *Astrocaryum* nuts, required an exceptionally long handling time (cf. Chapter 5).

Saimiri sciureus. Seasonal variation in the time budgets of *Saimiri* troops was minimal. They did rest less during the two dry-season samples (May, August), but the difference was small. The September sample coincided with the hottest period of the year, and it was obvious then that the animals reduced their activity on the warmest afternoons. Otherwise their behavior was remarkably constant.

Saguinus imperator. Interpretation of the data for both *Saguinus* species is somewhat complicated by the fact that the study troops moved to new territories midway through the study. However, it is appropriate to compare the August 1975 *imperator* sample with the October 1976 sample, because both of these were taken in the same territory, a large one of more than 50 ha. In August 1975 the animals were using the entire area and traversing its long axis nearly every day. Travel time was accordingly high (33%), while feeding time was average. Relatively little foraging was recorded, but as this was our first experience in watching *Saguinus*, the data may not be comparable to those taken in later samples.

The situation we found in October 1976, when we next followed the troop, was radically different. The female was conspicuously near the end of a pregnancy and moving with evident difficulty and reluctance. She was unable to travel at a normal pace, and invariably brought up the rear of a progression. Fruit was more abundant than it had been in the August sample, or at least more concentrated, because the troop remained within a small portion of the territory that contained several closely spaced trees. The change in behavior could have been a concession to the temporary incapacity of the female, or merely a response to an increased density of feeding trees.

Saguinus fuscicollis. Of the five species, *S. fuscicollis* was in a class by itself in its propensity to rest (44%). This was especially true in the November sample, when again special conditions seemed to apply. The female had given birth to twins just a week or two before the sample began. The young were being carried by the male, adding 30% or more to his weight. Food was superabundant at the time (cf. Chapter 2), and many suitable trees were available within a small area. The behavior of the *fuscicollis* troop was similar to that of the *imperator* troop with the pregnant female: they traveled little, rested inordinately (60%), and remained within a portion of their territory that contained a high density of fruit trees. When we next observed them six months later in April 1977, the young were fully independent and the group was using a larger area and foraging more. Feeding times also varied quite appreciably among the samples. This was due to large differences in the handling times for certain resources (cf. Chapter 5).

Discussion and Summary

Time budgets offer valuable insights into how animals order their lives. Not only can the relative importance of different activities be judged, but the impact of seasonal alterations in the environment and in the food supply can be seen in a shifting emphasis on particular activities.

In comparing the time budgets of our five omnivores with that of a vegetarian species (*Callicebus*), the most conspicuous contrast was in the abundant time available to the vegetarian for rest. Within the whole set of species, a close reciprocity between foraging and rest brought out the major behavioral implication of insectivory: a crowded time budget with little slack for stepped-up effort in periods of scarcity. Body size is another consideration that bears heavily on the interaction between time budgets and diet. Large animals need to eat more but do not gain a proportional advantage through their size in improved assimilation rates. This is especially true as applied to the capture of arthropod prey, a fact that imposes severe upper limits on the sizes of insectivores. Among birds the maximum is around 200 g, and among mammals it appears to be similar. Omnivores can be larger, but not too much larger, if insects are to be the principal source of protein. At a weight of 3 kg, *Cebus* is probably near the upper limit.

Larger Old World monkeys such as macaques, mangabeys, and guenons have been reported to search for arthropod prey, but the accompanying behavioral observations are in most cases scanty and not easily compared with ours. The long-tailed macaque is reported as spending 3% of its time on animal matter (Aldrich-Blake 1980). Low levels of insectivory in the mangabey (*Cercocebus albigena*) are indicated in Waser's (1977) results which show about 10% of the total time budget being devoted to scanning and manipulation of substrates in search of invertebrates. Similarly, the blue monkey (*Cercopithecus mitis*), searches a wide range of substrates, but the recorded average rate of prey capture was only 0.7 per hour (Rudran 1978). In Chapter 6 we shall show that the capture rates of *Saimiri* and *Cebus* are nearly two orders of magnitude faster. Nevertheless, some Old World monkeys do obtain appreciable amounts of arthropod prey, as shown by the stomach content data reported by Gautier-Hion (1980) for three species of guenons, sympatric in the forests near Makokou, Gabon (*Cercopithecus nictitans, C. cephus,* and *C. pogonias*). All contained 10–15% animal matter by dry weight, but corresponding time budgets are not available. In any case, at average body weights of 4–5 kg, the animals are only slightly larger than *Cebus*.

An interaction between size and insectivory is apparent in the fact that *Cebus* and *Saimiri*, the larger species, devote considerably more time to foraging than do the tamarins. Later, in Chapter 6, we shall see that the tamarins probably harvest as many grams of prey per hour as *Cebus*. At roughly one-tenth the body weight, it is obvious that the tamarins need to forage less to achieve the same relative protein intake.

Daily activity cycles also appear to be molded by energetic considerations. A charge of calories is needed to launch the day, and another one to carry the animals through the night. In this, the patterns shown by all the species are alike. There is a peak of fruit eating in the early morning and another in the late afternoon. Foraging is an interstitial activity at all hours, but is most concentrated in the middle of the day. Rest, quite reasonably, is most frequent and prolonged when the daily temperature cycle is near its peak.

Seasonal deviations from the mean time budget are interpreted less easily because of at least three complicating factors: major fluctuations in the quantity and quality of available resources, coincidence of the peak occurrences of births with the period of maximum fruit abundance, and a day-length cycle of appreciable amplitude (ca. 1.5 hr.). All species behave according to expectation in resting more during the long-day half of the year, but whether the behavioral shift in any individual case is a response to extra day length *per se* cannot be determined. Almost every sample had its idiosyncrasies. In some instances there were major changes in activity patterns accompanying the impending or actual arrival of babies in the troop (*Saguinus*); in others heavy rainfall was a disturbing factor (*C. albifrons*); and in still others seasonal dietary shifts called for major adjustments in the time allocated to feeding (*Saguinus, C. albifrons*). While serving to illustrate the multiplicity of influences that can bear on time budgets, these special effects resulted only in minor overlays on patterns established by the major causal factors of body size and dietary mode.

5 The Use of Plant Resources

We saw in the previous chapter that all five species of monkeys devote major amounts of time to feeding on plant resources and to foraging for prey. The differences between plant and animal resources in their degrees of dispersion and in the techniques used to harvest them are so profound that it seems best to discuss these two components of the diet separately. This I do in the present chapter and in the one that follows.

Ideally, one would like to be able to evaluate, in caloric or nutritional terms, the relative importance of each component of a species' diet. Simple, time-dependent measures of use are certain to contain distortions. Ingestion of concentrated plant resources, for example, is undoubtedly much faster per unit time than ingestion of dispersed arthropods. Even among different classes of plant products, e.g., fruits vs. leaves, rates of ingestion by primates are known to differ substantially (Hladik 1977). Feeding rates on a given resource may also vary considerably through time, for instance, as the crop of fruits on a tree becomes thinned out during the course of exploitation. Controlling all the important variables, and carrying out the myriad nutritional analyses that would have been required to put the diets of the five species on a quantitative footing, would have been another major project in itself. As our objective was to make broad general comparisons between the species, pursuing the issue of diet in such detail would have meant sacrificing other aspects of the study that were deemed more important. We thus resigned ourselves, as had most other primate workers before us, to living with the imperfections of time-dependent feeding data. The figures we present are thus not to be taken at face value as representing the importance of different components of the diet. They do, however, provide a basis for making broad comparisons between the species and between seasons.

The chapter begins by presenting data on the overall composition and diversity of plant materials consumed by the five species. Then, in an effort to distinguish which features of fruits and fruit trees are important in resource partitioning in the species under consideration, I examine a number of variables one at a time (e.g., fruit size, color, tree height, etc.). Finally, in the last section, I stress the adaptive interplay between diet, especially as constituted during the early dry season period of fruit scarcity, and divergent aspects of the behavior and morphology of the species.

Composition and Diversity of Plant Resources Consumed by Cocha Cashu Primates

Types of Plant Resources

The composition of the diets of all the species studied is strongly dependent on the availability of fleshy fruits. During most of the rainy season when such fruits are in ample supply, the monkeys converge in their feeding habits, concentrating on this single resource category to the extent of 96–100% (Table 5.1). Other categories of plant materials, though often no less available at other times of year, are used mainly or exclusively during the dry season when fruit is scarce. This strongly suggests that fruit is the preferred resource whenever it is available.

From May to September the species diverge in conspicuous ways. The *Cebus* species shift to palm nuts and pith, with a minor supplement of nectar. The *Saguinus* concentrate heavily on nectar and visit sap trees more often than at other times of the year. Only *Saimiri* succeeds in retaining a high percentage of fruit in its diet. We shall examine the dry season feeding behavior of all the species in greater detail in the latter part of the chapter. The important point to make here is that their diets are much less alike when fruit is scarce than when it is abundant.

Diversity of Plant Resources

The number and diversity of plant materials in the diets of the five primates fluctuate in a regular seasonal cycle that roughly parallels the fruiting cycle of the forest (Table 5.2). The changes are particularly dramatic in the two *Cebus* species and in *Saimiri*. During the portion of the year that extends from August or September through March, they eat a diversified assortment of ripe fruits, tasting or sampling casually many more species than are consumed in quantity.

The situation changes abruptly at the end of the rainy season. For the next three or four months (May through July or August), the number of available resources is drastically reduced and the animals are obliged to concentrate on just a few key plant species. Notwithstanding the limited array of choices, divergence in feeding habits is much more conspicuous than it is during the rainy season when the animals can select from a wide range of alternatives.

Such changes are apparent, though much less pronounced, in the record for tamarins. Their diets tend to be far less diversified than those of the larger monkeys at all times of year. This results from their singular habit of exploiting intensively just one or two plant species at a time, regardless of how many species may be fruiting in their territories. The only departure from this rule

Table 5.1
Plant materials eaten during the wet (Oct.–April) and dry (May–Sept.) seasons by Cocha Cashu primates.

Species	No. samples	Fruit		Seeds		Nectar		Sap		Fungi		Pith		Meristem		Flowers		Petioles		Leaves	
		W	*D*	*W*	*D*	*W*	*D*	*W*	*D*	*W*	*D*	*W*	*D*	*W*	*D*	*W*	*D*	*W*	*D*	*W*	*D*
Cebus apella	3W, 2D	99	66		25		1					<1	7		1			<1			
Cebus albifrons	2W, 2D	99	53		42		3						<1	<1	1				<1		
Saimiri sciureus	2W, 2D	100	91				9														
Saguinus imperator	1W, 2D	97	41			1	52	1	2	1							4				
Saguinus fuscicollis	2W, 1D	96	16				75	3	9		<1							1			

NOTE: Percentage of time spent feeding on plant resources averaged over all samples in each season.

Table 5.2
Number and diversity of plant resources used in seasonal samples.

Species	Season: Early rain (Oct.–Dec.)		Late rain (Jan.–Apr.)		Early dry (May–July)		Late dry (Aug.–Sept.)	
Cebus apella	17	(9.9)	18	(6.1)	8	(4.3)	14	(4.6)
Cebus albifrons	14	(6.8)	17	(11.7)	3	(2.4)	8	(2.0)
Saimiri sciureus	9	(5.1)	14	(7.8)	6	(1.9)	11	(6.6)
Saguinus imperator	4	(3.3)	3	(1.2)	4	(1.7)	11	(6.0)
Saguinus fuscicollis	9	(2.5)	5	(1.3)	6	(1.7)		
Callicebus moloch[a]					5	(2.2)		

NOTE: Number of items contributing 1% or more to total feeding time. Unlike Table 5.1, individual plant species are counted here. Diversity (in parentheses) computed as $\frac{1}{\Sigma p_i^2}$, where p_i represents the proportion of feeding time on the ith item in the diet.
[a] Fruit only; leaves and flowers mostly not identified.

was the August 1975 sample on *S. imperator*. In the light of what we now know, it is clear that this was a transitional period in which animals were switching their attentions from one set of resources to another. We shall return later to the novel feeding habits of *Saguinus*.

Taxonomic Composition

Over the course of a year, the species studied ate the fruit or other parts of over 170 species of plants in 55 families (Table 5.3). This can be compared to the presently known flora of about 1,000 species at Cocha Cashu. Perhaps 20% of this total is used in some way by the five species of monkeys and an unknown proportion of the remaining species are used by the other primates in the community.

Among the better represented families in the local flora, the most important in producing fruit for monkeys are the Moraceae, Annonaceae, and Palmae. Within a vast array of legumes, only the genus *Inga* is of major significance. The Lauraceae, with at least 14 species in the study area, produces fruits of appropriate size and construction (drupes) which are all but ignored by our monkeys, although they are eaten by *Ateles, Callicebus*, and many birds. The plants in other large families are either too small in stature to be easily used by primates (Melastomaceae, Piperaceae, Rubiaceae, Solanaceae) or produce fruits of unsuitable character (dry capsules, pods, samaras, etc.: Bignoniaceae,

Table 5.3

Characteristics of plant foods eaten by primates at Cocha Cashu.

Species[a]	*Life form*[b]	*Mean*[c] *ht(m)*	*crown dia(m)*	*Part eaten*[d]	*Type of fruit*[e]	*Dimensions of fruit (cm)*[f]	*Color when ripe*[g]	*Period used*	*Eaten by*[h]
Acanthaceae									
Mendoncia hirsuta	V?			fr	drupe	2×1	purple	Feb	Ss
Anacardiaceae									
Spondias mombin	T	(33)	(19)	fr	drupe	3×2.5	yellow	Jan-Feb	Cap,Cal,Ss
Tapirira guianensis	T			fr	drupe	1.5×1.0	pur-black	Feb	Si
Anonaceae									
Annona neglecta	T			fr	berry	2.5×2.0	yellow	Apr	Si
Duguetia quitarensis	T	10	7	fr	hrd berry	8.5	red	Jan-May	Cap,Cal,Ss,Si,Sf
Guatteria sp. 1	T	23	13	fr	drupe	2.0×1.0	dk purple	Jul-Oct	Cap,Ss
Guatteria sp. 2	T			fr	drupe	2.0×1.0	dk purple	Sep-Dec	Cap,Cal,Ss,Si,Sf,Cm
Malmea dielsiana	T			fr	drupe	2.0×1.0	?	Apr	Si,Sf
Malmea aff. *lucida*	T	(20)	(35)	fr	drupe		or-red	May-Jul	Cal,Ss
Malmea dichina	T			fr	drupe		red	Aug	Cap
Oxandra acuminata	T	11	5	fr	drupe	2.0×1.5	dk purple	Jan-Mar	Cap,Ss
Oxandra espintana	T			fr	drupe	1.5×1.0	red	Nov	Cap,Ss
Oxandra polyantha	T			fr	drupe	2.0×1.5	red	Nov	Ss
Unonopsis matthewsii	T	(25)	(15)	fr	drupe		red	Dec	Cal
Xylopia ligustrifolia	T	18	8	fr	capsule	1.5	pink	Feb-Mar	Cal,Ss
Apocynaceae									
Rauvolfia praecox	T				berry			Aug-Oct	Cap,Cal,Ss
Bonafousia sananho	T			fr	berry	3.5	yellow	Aug	Cap

(Table 5.3 cont.)

Species[a]	Life form[b]	Mean[c] ht(m)	Mean[c] crown dia(m)	Part eaten[d]	Type of fruit[e]	Dimensions of fruit (cm)[f]	Color when ripe[g]	Period used	Eaten by[h]
Araceae									
Anthurium kunthii	TC			fr	berry	0.5 × 0.5	red	Dec	Cal
Heteropsis oblongifolia	TC			fr	berry	1.0 × 1.0	red	Jan	Cap
Syngonium podophyllum	TC			fr	berry		yel-or	Mar-Apr	Cap,Cal,Si
Undet. sp.	TC			fr	berry			Oct-Nov	Cal,Sf
Undet. sp.	TC			pet	berry			May-Jun	Cap,Cal,Sf,Cm
Bignoniaceae									
Adenocalymma impressum	V			fl				May	Cm
Bombacaceae									
Quararibea cordata	T	26	12	fr	berry	6 × 4	orange	Jan-Mar	Cap,Cal
Quararibea cordata				nec				Aug-Sep	Cap,Cal,Ss,Si,Sf
Quararibea rhombifolia	T			fr	berry		orange	Dec-Feb	Cap,Cal
Quararibea witti	T	(15)	(7)	fr	berry		orange	Nov-Jul	Cap,Cal,Sf
Burseraceae									
Protium neglectum	T	(11)	6	fr	capsule	2.0	red & white	Feb	Ss
Protium tenuifolium	T			fr	capsule	2.5 × 2.0	red & white	Aug	Si,Sf
Cactaceae									
Epiphyllum phyllanthus	E			fr	berry		pink	Dec	Cal
Capparidaceae									
Capparis sola	T			fr	capsule	5.0 × 3.0	maroon	Jul	Ss,Cm
Caricaceae									
Carica microcarpa ssp. *heterophylla*	T	(4)	(1)	fr	berry	2.5	orange-red	Mar-Apr	Si
Jacaratia digitata	T	23	(8)	fr	berry	8.5 × 2.0	yellow	Jan-Apr	Cap

Celastraceae									
Maytenus magnifolia	T	(9)	(4)	fr	capsule	2.0	yellow	Jun-Dec	Cal,Ss
Chrysobalanaceae									
Hirtella racemosa	T			fr	drupe	1.5 × 1.0	purple	Jul-Oct	Ss
Licania britteniana	T			fr	drupe	2.5 × 1.5	yellow		Ss
Combretaceae									
Combretum assimile	V	28	8	nec				Jul-Aug	Cap,Cal,Ss,Si, Sf,At,Ceb
Terminalia oblonga	T			sap				Jul	Sf
Convolvulaceae									
Maripa cf. *peruviana*	V	(25)	(9)	fr	drupe	2.5 × 1.5		Sep	Cap,Ss
Dicranostyles ampla	V			fr	drupe	3.0 × 2.0	dk purple	Oct-Nov	Cap,Ss
Cucurbitaceae									
Cayaponia sp?	V			fr	berry	> 8.0	orange	Jun	Cal
Cyclanthaceae									
Undet.				pith				Jun	Cal
Dilleniaceae									
Doliocarpus dentatus	V			fr	capsule	1.5 × 1.0	red & white	Sep	Ss
Ebenaceae									
Diospyros cf. *pavonii*	T	(22)	(10)	fr	berry	2.5	dk red	Jul-Aug	Ss
Diospyros subrotata	T	(9)	(5)	fr	berry	3.5	yellow	May	Cap,Ss
Diospyros sp.	T			fr	berry	2.5		Jul-Sep	Cap,Ss
Elaeocarpaceae									
Sloanea guianensis	T	20	14	fr	capsule	2.0 × 1.5	red & white	Jan-Apr	Cap,Cal,Ss,Al,Ceb
Sloanea cf. *obtusifolia*	T	33	25	fr	capsule	2.5 × 2.0	red & white	Feb-Apr	Cap,Cal,Ss
Sloanea sp.	T			sap				Jul	Sf

(Table 5.3 cont.)

Species[a]	Life form[b]	Mean[c] ht(m)	crown dia(m)	Part eaten[d]	Type of fruit[e]	Dimensions of fruit (cm)[f]	Color when ripe[g]	Period used	Eaten by[h]
Euphorbiaceae									
Richeria cf. *racemosa*	T			fr	capsule	1.5	yel & red	Feb	Cap
Sapium aereum	T	(22)	(12)	fr	capsule	1.5×1.0	purple & red	Jan-Feb	Cap
Flacourtiaceae									
Casearia decandra	T	18	4	fr	capsule	2.5	white	Feb-Apr	Cap,Cal,Ss,Si,At
Casearia fasciculata	T	(7)	(6)	fr	capsule	3.5	yellow	Dec-Jan	Cap
Casearia sp. 1	T			fr	capsule	1.5	yellow	Feb-Mar	Ss
Casearia sp. 2	T			fr	capsule	2.0×1.5	or-red	Feb-Mar	Ss
Lunania parviflora	T			fr	berry		purple	Sep-Oct	Ss
Mayna parviflora	T	(7)	(6)	fr	hrd berry	4.0	yellow	Jan-Mar	Cap
Gramineae									
Bamboo sp.	C			lvs				May	Cm
Gynerium sagittatum	C			pith				Apr, Jun	Cap
Guttiferae									
Calophyllum brasiliense	T			fr	drupe	2.5	green	Feb-Mar	Cal
Chrysochlamys sp.	T			nec	capsule		red & white	Apr	Si
Rheedia acuminata	T	12	5	fr	berry	4.0	yellow	Oct-Dec	Cap,Cal
Rheedia sp.	T				berry	4.0×3.0	yellow	Sep-Dec	Ss
Icacinaceae									
Calatola venezuelana	T			fr	drupe	7.5×5.0		Jun	Cal
Calatola sp. nov.	T			fr	drupe	4.0×2.5	yellow	May-Jul	Cm,At
Leretia cordata	TC			fr	drupe	2.5×2.0	dk red-pur	Apr-Aug	Ss,Sf
Lauraceae									
Undet. spp.	T	20	9	fr	drupe	1.5×1.0	black	Sep-Dec	Cap,Cal,Ss,At,Cm

Leguminosae									
Cesalpinia bonduc	V	12	6	seed	capsule	9.5 × 6.5	brown	Jun-Aug	Cap,Ss
Copaifera sp.	T			fr	capsule		brn & yel	Aug	At
Inga cf. *chartaceae*	T	17	10	fr	hrd berry	29.0 × 4.0	green	Jan-Feb	Cap
Inga edulis	T	18	14	fr	hrd berry	24.0 × 2.0	green	Jan-Mar	Cap,Cal,Ss
Inga edulis				sap					Sf
Inga marginata	T	15	10	fr	hrd berry	12.0 × 2.0	green	Jan-Apr	Cap,Cal,Ss,Si
Inga mathewsiana	T	14	8	fr	hrd berry	13.0 × 3.5	green	Oct-Apr	Cap,Cal,Ss,Sf
Inga mathewsiana				sap					Si,Sf,Ceb
Inga nobilis	T	24	14	fr	hrd berry	18.0 × 3.0	green	Aug-Feb	Cap,Ss,Al
Inga pavoniana	T	(20)	(12)	fr	hrd berry		green	Jan	Cap
Inga punctata	T	17	8	tr	hrd berry	34.0 × 2.5	green	Jan-Mar	Cap,Cal,Ss
Inga cf. *splendens*	T	17	7	fr	hrd berry	11.0 × 4.0	green	Jan-Feb	Cap
Inga sp.	T			sap				Apr, Nov	Si,Sf
Lecointea peruviana	T	17	9	fr	berry	5.0 × 2.5	white	Nov-Dec	Cal,Ss
Loganiaceae									
Strychnos asperula	T	30	7	fr	hrd berry	3.5	yellow	Aug-Nov	Cap,Ss,Al,At
Malpighiaceae									
Banisteria sp.	V			fr	samara	3.0 × 1.5	brown	Sep	Cap
Malvaceae									
Malvaviscus sp.	V	9	4	fr	berry		white	Jul-Aug	Si,Sf
Marcgraviaceae									
Marcgravia macrocarpa	TC			fr	capsule	3.0 × 2.5	brown & red	Nov-Dec	Cal
Marantaceae									
Calathea sp.	H			fr	capsule		white	Apr	Cap
Ischnosiphon puberulus	H			pith				Jun	Cap

(Table 5.3 cont.)

Species[a]	*Life form*[b]	*Mean*[c] *ht(m)*	*Mean*[c] *crown dia(m)*	*Part eaten*[d]	*Type of fruit*[e]	*Dimensions of fruit (cm)*[f]	*Color when ripe*[g]	*Period used*	*Eaten by*[h]
Melastomataceae									
Miconia splendens	T	(10)	(3)	fr	berry	0.5	dk purple	Jun-Jul	Cap,Ss
Undet.	T			fr	berry			Jun	Cal
Meliaceae									
Cedrela odorata	T			sap				Jul	Si,Sf
Guarea sp.	T	(9)	(3)	fr	capsule		red & brown	Apr	Si
Trichilia pleeana	T	15	10	fr	capsule	2.0 × 1.5	green & red	Jan-Feb	Cap,Ss
Trichilia poeppigii	T	(14)	(5)	fr	capsule	3.0 × 2.0	white & red	Sep-Feb	Cap
Trichilia quadrijuga	T			fr	capsule	1.5 × 0.5	or & red	Nov	Sf
Trichilia sp.	T			fr	capsule		red	Sep, Dec	Cal,Ss
Menispermaceae									
Anomospermum chloranthum	V			fr	drupe	3.5 × 3.0	orange	Mar-May	Ss
Anomospermum grandifolium	V	(35)	(18)	fr	drupe	4.5 × 3.5	orange	Sep-Feb	Cap
Anomospermum reticulatum	V	12	6	fr	drupe	2.5	orange	May-Jul	Cap,At
Chondrodendron tomentosum	V			fr	drupe	2.5 × 1.0	dk purple	Mar-Aug	Ss,Si,Sf
Curarea toxicofera	V			fr	drupe	2.5 × 1.0	yel-or	Aug	Si,Sf
Odontocarya arifolia	V			fr	drupe	1.5 × 1.0	yellow	Apr-Aug	Si,Sf
Borismene japurensis	V			fr	drupe		red	Sep	Ss

Moraceae									
Brosimum alicastrum	T	30	23	fr	drupe	2.0	yellow	Jan-Mar	Cap,Cal,Ss,Al,At
Brosimum lactescens	T	40	25	fr	drupe	2.0	orange	May-Dec	Cap,Cal,Ss,Si,Sf,Al,At
Cecropia aff. *leucophaia*	T	18	11	fr	berry		brown	Jan-May	Cap,Cal,Ss,Sf,At
Clarisia racemosa	T	40	14	fr	drupe	2.5 × 1.5	red	Oct-Dec	Cap,Cal,Ss
Coussapoa sp.	E			fr	berry		brown?	Dec	Cal
Ficus cf. *amazonica*	T?	(23)	(8)	fr	berry	2.0		Oct-Nov	Cap,Cal,Ss
Ficus cf. *casapiensis*	E	16	6	fr	berry	1.5	red-purple	Mar-Oct	Cap,Cal,Ss,Si
Ficus erythrosticta	E	20	12	fr	berry	1.0	yel & red	all year	Cap,Cal,Ss,Si,Cm
Ficus cf. *expansa*	T	34	25	fr	berry	3.0	green	Apr-Dec	Cap,Cal,Ss,Sf,Al,At
Ficus insipida	T	30	23	fr	berry	2.5	grn-brown	Feb-May	Cap,Cal,Ss
Ficus killipii	T	30	25	fr	berry	1.0	red	Jul-Mar	Cap,Cal,Ss,Si,Sf,Al,At
Ficus mathewsii	T			fr	berry	0.5	yel-or	Aug-Sep	Cap,Ss,Al,At
Ficus maxima	T	20	14	fr	berry	2.5	green	Jan-Mar	Cap,Cal
Ficus paraensis	E			fr	berry	2.5 × 2.0	yel & purple	Aug-Sep	Ss
Ficus perforata	T	50	45	fr	berry	0.5	red	May-Oct	Cap,Cal,Ss,Cm,Al,At
Ficus pertusa	E	(30)	(24)	fr	berry	1.5	yel & red	Jun-Nov	Cal,Ss
Ficus cf. *regularis*	?			fr	berry	2.5	green	May	Cap,Ss
Ficus cf. *sanguinosa*	?			fr	berry	3.0	brown	May-Jun	Cap
Ficus trigona	T	15	25	fr	berry	1.5	red	Aug-Mar	Cap,Cal,Ss,Si,Cm,Al
Ficus ypsilophlebia	E			fr	berry	3.0	green	Aug	Ss
Ficus sp. 1	T	(37)	(40)	fr	berry	2.5	green	May	Cap
Perebea guianensis	T			fr	berry	1.5	red	Feb-Mar	Ss
Pourouma cecropiafolia	T	23	7	fr	drupe	2.5 × 2.0	dk purple	Sep-Mar	Cal,Ss,Si,Sf
Pourouma cf. *minor*	T			fr	drupe	3.0 × 2.0		Aug-Sep	Cap
Pourouma sp. 1	T			fr	drupe	2.0 × 1.5		Aug	Si,Sf
Pourouma sp. 2	T			fr	drupe	2.5 × 2.0		Aug-Oct	Cap,Ss
Pseudolmedia laevis	T	25	10	fr	drupe	2.0	red	Jul-Feb	Cap,Ss,Al
Sorocea cf. *briquetii*	T	17	9	fr	drupe	1.5 × 1.0	purple	Nov-Feb	Cap,Cal,Ss,Si,Sf

(Table 5.3 cont.)

Species[a]	Life form[b]	Mean[c] ht(m)	Mean[c] crown dia(m)	Part eaten[d]	Type of fruit[e]	Dimensions of fruit (cm)[f]	Color when ripe[g]	Period used	Eaten by[h]
Musaceae									
Heliconia sp.	H			pith				Dec, Jun	Cal
Myristicaceae									
Otoba parvifolia	T			fr	capsule	2.0	green & red	Jan	Cap
Myrsinaceae									
Ardisia nigrovirens	T			fr	drupe	1.0	dk purple	Jul-Aug	Ss
Myrtaceae									
Calyptranthes longifolia	T	(6)	(4)	fr	berry	2.5	yellow	Sep	Ss
Calyptranthes cf. *macrophylla*	T			fr	berry	2.5 × 2.0	purple	Jul-Dec	Ss
Calyptranthes cf. *multiflora*	T			fr	berry	2.0	purple	May	Si
Eugenia cf. *acrensis*	T	(5)	(3)	fr	drupe	2.5 × 1.0	orange	Oct-Nov	Cap,Si,Sf
Eugenia cf. *punicifolia*	T	(6)	(6)	fr	drupe	1.0	purple	Jul-Oct	Cap,Ss,Sf
Myrcia splendens	T	(9)	(6)	fr	drupe	1.5 × 1.0	purple	Feb-Mar	Ss
Myrciaria amazonica	T				drupe		dk blue	Dec	
Nyctaginaceae									
Neea chlorantha	T	9	4	fr	drupe		purple	Oct-Nov	Cap,Ss,Sf
Palmae									
Astrocaryum sp.	T	9	8	mer				Aug-Oct	Cap,Ss
Astrocaryum sp.	T	9	8	fr	drupe	4.0 × 2.5	orange	Jan-Apr	Cap,Cal
Astrocaryum sp.	T	9	8	seed	nut	4.0 × 2.5		Mar-Jul	Cap,Cal
Astrocaryum sp.	T	9	8	pith				Jun	Cap,Cal

Iriartia ventricosa	T	(18)	(7)	fr	drupe	1.5	black	Apr-Jun	Cap,Si,Sf,Cm
Iriartia ventricosa	T	(18)	(7)	sap				Apr, Jul	Si,Sf
Scheelea sp.	T	11	8	mer				Aug-Sep	Cap
Scheelea sp.	T	11	8	pith				Jun-Sep	Cap
Scheelea sp.	T	11	8	fr	drupe	8.0 × 3.0	brown	all year	Cap,Cal,Ss
Hyospathe sp.	T			fr	drupe		black	Apr	Cap
Hyospathe sp.	T			mer				Jun	Cap
Passifloraceae									
Undet.	V			fr	berry			Mar	Cal
Phytolaccaceae									
Trichostigma octandrum	V			fr	drupe	0.5	dk purple	Oct-Dec	Cap,Cal,Ss,Sf
Quiinaceae									
Quiina peruviana	T	(26)	(7)	fr	drupe	3.5 × 2.5	orange	Dec	Cal
Rubiaceae									
Alibertia pilosa	T			fr	berry		yellow	Nov	Sf
Genipa americana	T			fr	hrd berry	8.5 × 6.0	brown	Dec	Cal
Randia sp. 1	T			fr	berry	2.5 × 2.0	yel-or	Apr	Si
Randia sp. 2	T			fr	berry	2.5	orange	Jul	Ss,Sf
Sapindaceae									
Allophyllus glabratus	T	(11)	(5)	fr	drupe	1.0	yel-or	May	Ss
Allophyllus scrobiculatus	T	15	8	fr	drupe	0.5	yellow	Apr-Jul	Cap,Ss,At
Paullinia alata	V			fr	capsule		red & white	Mar	Ss
Paullinia capreolata	V			fr	capsule	2.0 × 1.5	yel & white	Feb-Mar	Cap,Ss
Paullinia hystrix	V	21	6	fr	capsule	3.5 × 3.0	yel-or & white	Oct-Dec	Cap,Cal,Ss
Paullinia obovata	V	(19)	(3)	fr	capsule		yel & white	Mar-Apr	Ss,Si
Paullinia sp.	V				capsule			Mar	Cal
Undet.	V			lvs				May	Cm

(*Table 5.3 cont.*)

Species[a]	Life form[b]	Mean[c] ht(m)	Mean[c] crown dia(m)	Part eaten[d]	Type of fruit[e]	Dimensions of fruit (cm)[f]	Color when ripe[g]	Period used	Eaten by[h]
Sapotaceae									
Manilkara sp.	T	(33)	(20)	fr	berry	2.5 × 2.0	orange	Jan-Feb	Cap,At
Pouteria cf. *anibaefolia*	T	(8)	(6)	fr	berry	2.5 × 1.5	yellow	Jan-Feb	Cap
Pouteria cf. *boliviana*	T	(25)	(14)	fr	hrd berry	4.0 × 3.5	yellow	Jan-Jun	Cap
Pouteria cf. *cylindrocarpa*	T			fr	berry	3.5 × 3.0	yellow	Sep-Oct	Ss,Al
Pouteria cf. *ephedrantha*	T			fr	hrd berry	10.0 × 7.0	brown	Mar	Cal
Pouteria cf. *nitida*	T			fr	berry	7.0 × 6.0	orange	Aug	At
Pouteria cf. *ulei*	T			fr	berry	3.5 × 2.5	yellow	Mar	Cal
Pouteria cf. *venulosa*	T	26	16	fr	berry	3.0 × 2.0	purple	Jan-Mar	Cap,Ss
Pouteria sp. 1	T			fr	berry	3.5	yellow	Oct-Nov	Cap
Simaroubaceae									
Picramnia sp.	T			sap	drupe		red	Jul	Sf
Sterculiaceae									
Theobroma cacao	T	12	6	fr	hrd berry	18.0 × 10.0	yellow	Dec-Apr	Cap,Cal
Tiliaceae									
Apeiba membranacea	T	(18)	(5)	fr	hrd berry		black		Cap
Theophrastaceae									
Clavija tarapotana	T			fr	berry		yellow	Dec	Cal
Ulmaceae									
Ampelocera sp.	T	20	15	fr	drupe	1.0	yellow	Nov-Dec	Cal,Ss,At
Celtis iguanea	V	20	8	fr	drupe	1.5 × 1.0	yellow	Mar-Aug	Cap,Cal,Ss,Si,Sf, Cm,Al,Ao,At

Urticaceae									
Urera caracasana	T	8	6	fr	berry	0.1	yellow	Nov-Dec	Cap,Cal,Ss
Urera eggersii	V			fr	berry		orange	May-Jun	Cap,Ss
Violaceae									
Gloeospermum sphaerocarpum	T			fr	berry	3.0	yellow	Apr	Sf
Leonia glycycarpa	T	12	4	fr	hrd berry	4.5 × 4.0	white-grn	Jun-Aug	Cap,Cal,Ss,Si,Sf
Vitaceae									
Cissus sicyoides	V			fr	drupe	1.0	dk purple	Nov-Dec	Cal
Cissus ulmifolia	V	(14)	(3)	fr	drupe	1.5 × 1.0	purple	Aug-Sep	Cap,Cal,Ss,Si,Sf

[a] Species names follow those in MacBride (1936) whenever possible.

[b] C = cane; E = epiphyte or hemi-epiphyte; H = herb; T = tree; TC = trunk climber; V = vine.

[c] Mean height and crown diameter of plants used by monkeys; values in parentheses when sample size $\leqslant 3$.

[d] Plant parts: fl = flowers; fr = fruit; lvs = leaves; mer = meristem; nec = nectar; pet = petiole; pith; sap; seeds.

[e] Fruit types operationally defined as follows: berry—soft fruits with >1 seed, the seeds often small and swallowed intact; capsule—dihiscent fruits, usually with >1 seed encased in hard outer shell, often open to expose bright aril, the seeds usually not swallowed; drupe—soft, single-seeded fruits, the seeds usually ejected during mastication; samara—dry indehiscent fruits bearing a broad appendage(s) for wind dispersal; seed—recorded only when the seed was the part eaten after removal of outer layers of fruit.

[f] Approximate mean dimensions of long and short fruit axes; spherical fruits represented by diameter only.

[g] Normally the color of the ripe fruit exterior; two colors are listed for some striped and spotted fruits and for some arilate fruits when the color of the outer capsule and aril differ.

[h] Al = *Alouatta*; Ao = *Aotus*; At = *Ateles*; Cal = *Cebus albifrons*; Cap = *Cebus apella*; Ceb = *Cebuella*; Cm = *Callicebus moloch*; Sf = *Saguinus fuscicollis*; Si = *Saguinus imperator*; Ss = *Saimiri sciureus*.

Leguminosae, Malpighiaceae), or are both small in stature and produce unsuitable fruits (Acanthaceae, Araceae, Compositae, Gramineae). In contrast, the fruits of a few small families are conspicuously overrepresented in primate diets: Elaeocarpaceae, Flacourtiaceae, Loganiaceae, Ulmaceae, Urticaceae, Vitaceae. Still other families are widely used but in relatively small quantities: Ebenaceae, Guttiferae, Meliaceae, Menispermaceae, Myrtaceae, Sapindaceae, Sapotaceae.

It is not easy to find criteria that distinguish the species of fruit used by one primate from those used by others. The popularity of many important genera is nearly universal, e.g., *Ficus, Brosimum, Guatteria, Casearia, Inga, Cecropia, Celtis, Cissus*, etc. The fruits have no common color, size, or design except that they contain soft pulpy material surrounding the seeds. The most likely reason for their wide acceptance is that they are abundant and readily digestable. Most are, in fact, tolerable, if not exciting, to the human palate. The general impression one gains is that fruits which are eaten by one monkey are probably eaten by others as well, even though the observations are not extensive enough to prove the point. Gautier-Hion (1980) came to the same conclusion on the basis of more complete data for three sympatric species of *Cercopithecus* in Gabon. In our case, direct comparisons of dietary similarity are precluded by the sequential, rather than simultaneous ordering of samples. Nevertheless, in comparing successive samples on *C. apella, Saimiri*, and *C. albifrons* taken during February and March 1977, there is a wide overlap in the fruit species that rank high in the diets of all three species.

Properties of Fruits and Fruit Trees Important in Resource Partitioning

In this section I shall examine several of the obvious external characteristics of fruits, i.e., size, hardness, color, and show that these features are of relatively minor significance in discriminating the use of fruit species by the several species of primates. Of far greater importance are features of the plants that bear the fruits, especially their size.

Size of Fruits

In view of the fact that the five primates we studied are arrayed over a tenfold range in body size, one might expect to see a conspicuous degree of separation among them in the sizes of fruits eaten. Table 5.4 shows that there is a regular tendency for the larger species to eat larger fruits, but that the differences are small and purely statistical. All species ate fruits of all size classes (except

Table 5.4
Sizes of fruit eaten by Cocha Cashu primates
(percentage by feeding time, averaged over all samples).

	Fruit size (cm)[a]						
pecies	*0.0–0.5*	*>0.5–1.0*	*>1.0–2.0*	*>2.0–4.0*	*>4.0–8.0*	*>8.0*	*Weighted mean (cm)*[b]
ebus apella	8	20	31	37	2	2	1.6
Without palms	11	27	41	16	2	2	1.2
ebus albifrons	33	15	6	45	<1	1	1.1
Without palms	53	24	9	11	<1	2	0.6
aimiri sciureus	43	35	11	11	<1	<1	0.6
aguinus imperator	<1	87	7	1	4	<1	0.9
aguinus fuscicollis	1	93	1	3	<1	2	0.8
allicebus moloch	–	77	14	9	–	–	0.9

Diameter or short axis in the case of elongate fruits.
Logarithms of midpoints of class ranges used in calculating the weighted means.

for *Callicebus*, for which there was only one short sample). Surprisingly, *Saguinus* ate as much fruit in the largest two categories as *Cebus*, while *Cebus* ate far more fruit in the smallest category, mostly figs. Large fruits eaten by *Saguinus* were *Leonia glycycarpa* (July, August) and *Duguetia quitarensis* (April). Both *Cebus* species also ate *Duguetia* in small amounts, but for them the most important large fruit was *Theobroma cacao*. The *Saguinus* ate few figs, hence the paucity of tiny (<5 mm) fruits in their diets. *Saimiri* is at the opposite extreme in this respect, favoring figs over all other fruit. The concentration of its feeding in the smallest fruit categories results from the fact that the fruits of the three most important fig species (*Ficus perforata, F. killipii*, and *F. erythrosticta*) are all 1 cm in diameter or less.

Palm nuts account for much of the feeding of *Cebus* on fruits in the 2–4 cm class. When these are removed from the compilation, leaving mainly soft fruits, the differences between *Cebus* and the other species are greatly reduced. Fruit size is thus not a major factor in distinguishing the food habits of the five species.

Hardness of Fruits

The majority of fruits eaten by the monkeys were either soft to the touch or only lightly protected. Among the stouter, tougher fruits were the 10–30 cm long pods of the various *Inga* species, but some of these were used routinely, if not avidly, by *Saguinus*. A number of species may have posed difficulties for the smaller primates by being both large and tough (e.g., *Quararibea cordata, Anomospermum grandifolium, Genipa americana, Pouteria* spp.), but none of these was eaten in great quantity, and all of them ripened at times when many other alternative resources were available. The only large, tough fruit that was consumed regularly by *Cebus* was *Theobroma cacao*. The pods, some of them larger than an acorn squash, were too tough even for the *Cebus* to open directly. After biting or pulling them free from their attachments, the monkeys smashed them repeatedly against the branches until they broke. The prize was a mass of sweet pulp that surrounded the seeds in the central cavity of the fruit.

In general, we can conclude that fruit hardness is of little more importance than fruit size in prompting the differential use of plant species by the monkeys. The one outstanding exception to this occurs in the use of palm nuts. These are of the utmost importance to *Cebus* in the dry season. Gaining access to the nuts themselves or their contents is a task that requires the animals' full strength. The extensive use of nuts and other palm products is something that sets *Cebus* apart from all the other primates at the site.

Fruit Color

In the course of this work we became impressed by the consistency with which fruits eaten by the monkeys were of colors that fell into the yellow to orange region of the spectrum. Much later, when the appropriate data had been compiled, we found that our impression was verified (Table 5.5); yellow to orange fruits do indeed predominate by a wide margin. This becomes all the more remarkable when one thinks of the fruits of North American plants. The common colors are red, black, and blue to dark purple. Most such fruits are known to be bird dispersed (Stiles 1980). A few North American fruits are white (e.g., *Cornus* spp., *Myrica* spp.) or green (*Asimina, Maclura*); but none, to my knowledge, are bright orange or yellow. The closest approaches to these colors are found in some Rosaceous genera (*Prunus, Pyrus*, and *Crategus*) and in the persimmon (*Diospyros virginiana*). Few of these actually

Table 5.5
Color of fruits eaten by Cocha Cashu primates.

Number	Black	Purple-maroon	Green	Yellow-orange	Red	Brown	White
of species	7	17	16	55	29	8	5

NOTE: Color of fruit when fully ripe or, in some cases, of aril.

hit the mark, as the rosaceous fruits tend to be dull and either on the reddish or greenish sides, while the persimmon is more pinkish or salmon than orange.

How can we account for the extraordinary prevalence of yellows and oranges in the fruits eaten by primates? The likely answer is that the plants have adapted to the sensory capacities of the animals that serve as their principal dispersal agents. Old World monkeys have a three-pigment color vision system very much like our own. They see the full spectrum from violet to dark red (DeValois and Jacobs 1968). But New World monkeys possess a two-pigment system with a more limited spectral range. They are sensitive to wavelength differences in the blue, green, and yellow regions, but they do not discriminate red (DeValois and Jacobs 1971; Snodderly 1972). We may thus conjecture that to such a visual system, yellow and orange are the colors that stand out most conspicuously against a backdrop of green foliage.

Snodderly (1979) recently considered this issue, but came to the remarkably different conclusion that many of the fruits eaten by New World monkeys are cryptically colored. This impression was gained from observations on *Callicebus torquatus* in a comparatively impoverished white sand flora in northern Peru. Of the seventeen species of fruits described in his report, six were listed as being yellow, orange, or yellowish green, and four as green, when eaten by the monkeys. The color of the fruit or aril when fully ripe was noted for nine of the species, and in five of these cases it was yellow or orange. Snodderly's data thus do not seem to differ much from ours, but his conclusion does. We suspect that this is the result of his choice of *Callicebus torquatus* as the subject animal. His data indicate that the monkeys were eating large numbers of seeds from fruits that were not intrinsically cryptic, but merely immature. The seeds of mature fruits are likely to be hard and difficult to chew or digest, while immature seeds are soft and easily ingested. In eight out of nine species that were listed as being green or yellowish green when taken, the seed was the part eaten. *Callicebus torquatus* thus has a special reason for taking unripe fruit, and can be better regarded as a seed predator than as a dispersal agent. In contrast, the monkeys we studied, with few exceptions, distinctly prefer ripe fruits and in eating them do little or no damage to their seeds.

As with size and hardness, fruit color failed as a criterion for differentiating the use of fruit species by the five primates. Each was observed to consume fruits in every color category, and yellow to orange fruits were well represented in the diets of all of them.

HEIGHTS OF FRUIT TREES

Observers of Old World monkeys have often remarked that related species tend to feed at characteristically different heights in the forest (e.g., Booth 1956; Rodman 1978; Gautier-Hion 1980). This is true only to a limited degree in the five species we studied (Table 5.6). Overall differences between species tend to be little greater than the seasonal differences within species. *Saguinus* are the most consistent in their feeding heights (20–25 m), while the *Cebus* species show the greatest variability. Much of this in *Cebus apella* is due to seasonal alternation between palms (mostly <15 m) and large fruit trees (mostly >20 m). In *Cebus albifrons* the greatest contrast is between the March and June samples. In March the troops used large numbers of several common small species (*Inga* spp., *Theobroma cacao, Casearia decandra, Xylopia* sp., and *Cecropia leucophaia*), while in June feeding was highly concentrated in

Table 5.6
Percentage of feeding time vs. tree height.

		Tree height (m)				
Species	*Sample*	*0–10*	*11–20*	*21–30*	*>30*	*Weighted mean height*
Cebus apella	October	25	29	28	18	19
	January	1	36	24	39	25
	June	31	42	26	1	15
	Year	**21**	**44**	**20**	**15**	**18**
Cebus albifrons	November	19	35	27	19	20
	March	32	26	21	21	18
	June	8	5	14	73	30
	Year	**20**	**22**	**21**	**38**	**23**
Saimiri sciureus	October	0	8	18	73	32
	February	13	60	14	13	18
	May	7	15	4	74	30
	August	1	13	45	41	28
	Year	**5**	**24**	**20**	**50**	**27**
Saguinus imperator	April	2	42	51	4	21
	July	5	4	77	14	25
	Year	**4**	**23**	**64**	**9**	**23**
Saguinus fuscicollis	November	3	34	63	–	21
	April	4	48	41	7	20
	July	1	17	72	10	24
	Year	**3**	**33**	**59**	**6**	**22**

two large fig trees. *Saimiri* consistently fed higher than the other species, often in giant canopy emergents. Although the troops regularly foraged near or on the ground for insects, and individual monkeys often took fruits from low trees, major feeding bouts that engaged the whole group could only occur in large trees because of the number of animals involved. This trend is all the more pronounced as a result of the strong predilection *Saimiri* shows for figs. Many of the trees are stranglers that have overtopped their already tall hosts in the competition for full sunlight. Figs accounted for more than 50% of the feeding time of *Saimiri* in all but the February sample, when they were only a minor item (7%). The mean feeding height then shifted to a much more moderate level (18 m).

The *Saguinus* differ from the other species in their avoidance of the highest and lowest classes of trees and in their seasonal consistency. Part of the explanation for this is that much of their feeding takes place in vines. These are most concentrated in the subcanopy zone where they often form extensive tangles that tie together several crowns. Other important *Saguinus* resources are subcanopy trees. The animals rarely ascend to the highest levels of the canopy, seemingly because they are loath to leave the protective cover of sheltering vines and foliage.

None of the species find much fruit in the understory below 10 m, especially if one discounts the palms used by *Cebus*. The probable reason for this is that small plants growing in the shade are unable to produce crops of sufficient size to attract animals as large as monkeys. Many of the plants that do grow at this level have small berry-like fruits that are primarily bird-dispersed, e.g., many Melastomaceae, Myrsinaceae, Nyctaginaceae (*Neea* spp.), Rubiaceae, and Solanaceae. There may be a hindrance in the weak mechanical support offered by low trees and shrubs, but this cannot be the only explanation as all five species do a significant amount of insect foraging below 10 m.

Crown Diameters of Fruit Trees

Crown diameter is a much better indicator of overall tree size in the tropical forest than tree height. This is because many species grow to considerable heights before making a major investment in branching structure. Many mid-story and subcanopy trees possess slender, unbranched trunks with a narrow topknot of foliage. There are few opportunities to spread out laterally below the canopy itself. Crown diameters of more than 15 m are thus indicative of canopy trees—ones that have reached an open site and that enjoy full or nearly full sunlight. Not all such trees are dramatically tall, because the canopy of a tropical forest is an incredibly irregular patchwork of openings and trees of all sizes. The important thing about large trees for our purposes is that they are likely to be exposed to full sun and thus to be more productive per unit area than small-crowned trees that are still striving to reach the canopy.

As can be seen in Table 5.7, crown diameter serves to differentiate the behavior of the monkeys far better than did tree height. Now it emerges unambiguously that the *Saguinus* use small trees almost exclusively, while *Saimiri* concentrates its feeding in crowns of the largest diameter classes. The *Cebus* are more versatile in using trees over the full spectrum of sizes. The modal size for *Cebus apella* is small because of its concerted seasonal use of palms, and for this reason the variability between samples is high. *Cebus albifrons*, as mentioned previously, fed in numerous small to medium-sized

Table 5.7
Percentage feeding time vs. crown diameter.

		Crown Diameter (m)						*Weighted mean*
Species	*Sample*	*0–5*	*6–10*	*11–15*	*16–20*	*21–30*	*>30*	*diameter*
Cebus apella	October	7	65	7	14	7	–	11
	January	1	15	16	27	26	15	20
	June	5	59	20	15	–	–	10
	Year	**4**	**46**	**14**	**19**	**11**	**5**	**14**
Cebus albifrons	November	7	47	4	18	13	11	15
	March	21	33	20	2	24	–	12
	June	1	11	2	12	74	–	22
	Year	**10**	**30**	**9**	**11**	**37**	**4**	**17**
Saimiri sciureus	October	–	10	14	10	41	24	23
	February	12	30	26	11	10	11	14
	May	1	19	5	1	1	73	28
	August	5	27	19	4	36	9	18
	Year	**4**	**22**	**16**	**7**	**22**	**29**	**21**
Saguinus imperator	April	15	69	16	–	–	–	8
	July	3	58	39	–	–	–	10
	Year	**9**	**64**	**27**	**–**	**–**	**–**	**9**
Saguinus fuscicollis	November	9	82	9	–	–	–	8
	April	13	66	15	2	4	–	9
	July	14	52	34	–	–	–	9
	Year	**12**	**67**	**19**	**1**	**1**	**–**	**9**

trees in November and March, but in June logged the major part of its feeding time in just two large figs.

The criterion of crown size serves to affect a major degree of separation in the feeding activities of the five species. More than half the feeding of *Cebus apella* and *Saguinus* is done in trees with crowns less than 15 m in diameter, but the species in question are almost entirely different. For *apella* they are mainly palms plus a few dicots such as *Allophyllus scrobiculatus* (Sapindaceae), *Trichostigma octandrum* (Phytolaccaceae), *Pseudolmedia laevis* (Moraceae), and *Strychnos asperula* (Loganiaceae), none of which the *Saguinus* used at all. In turn, the *Saguinus* concentrated on plants that were little used by the other three species, as will be further explained in a later section of the chapter. *Saimiri* and *Cebus albifrons* separate from the others in doing more than half of their feeding in crowns larger than 15 m in diameter. Here the overlap is considerable because both species are heavy users of figs. But even when figs are not predominant in the diets, the overlap is high. This can be seen in the back-to-back February and March samples when the principal genera in the diets of both species were *Sloanea, Inga, Ficus, Casearia,* and *Cecropia.* The only species eaten extensively by *albifrons* (9%) that was not also important in the diet of *Saimiri* was *Theobroma cacao*, the pods of which are too tough and heavy for *Saimiri* to handle. It is thus clear that these two species are very close in their fruit-eating habits. This fact may not be of much importance for most of the year, but brings them into intense competition during the dry season when figs are almost the only available fruit.

Divergent Patterns in the Use of Fruit Resources

Use of Palms by *Cebus*

Cebus apella uses palms in an impressive number of ways: for insect foraging (all species), for the fleshy or semihard exo- or mesocarp (*Astrocaryum, Bactris, Iriartia, Mauritia, Scheelea*), for the seed itself (endosperm, *Astrocaryum*), for the pith contained in frond petioles (*Astrocaryum, Scheelea*), for the apical meristem (''heart of palm'': *Bactris*), and for immature inflorescences (*Astrocaryum, Scheelea*). We believe that the singular versatility of *Cebus apella* in being able to utilize the resources offered by palms, especially the abundant *Astrocaryum*, is what accounts for its relatively small home range (compared to *C. albifrons* and *Saimiri*: cf. Chapter 8).

During the annual period of fruit scarcity, palms provide a ubiquitous backup resource. A survey conducted by R. Kiltie (1980) found that the two principal species, *Astrocaryum* and *Scheelea*, occur at surprisingly uniform densities throughout virtually all habitats over the entire study area. The mean densities for the two species were, respectively, 39 and 25 per hectare. This means that the average troop in its 50 ha home range has access to more than a thousand trees of each species. The nuts of *Astrocaryum* mature in April and are abundantly available during the May–June period of greatest fruit scarcity. Later in the season (July, August) the supply of *Astrocaryum* nuts is gradually depleted through the combined depredations of bruchid beetles, *Cebus* monkeys, macaws, peccaries, squirrels, and other rodents. By this time, the future inflorescences are in an immature stage and *Cebus apella* switches to these. They consist of masses of soft material of presumably high nutritional value. Although well protected in the crown of the palm by a tough woody sheath and a formidable armament of needle-sharp spines, they are regularly extracted by the redoubtable animals. *Cebus apella* can also remove long cylinders of pith from the incredibly hard woody petioles of the 6 m long fronds. In these feats, *apella* has no competitors. Palms thus represent the primary ecological refuge of the species.

Neither species of *Cebus* used palm nuts or other palm products to any significant degree during the season of fruit abundance (October–March). This was perhaps for the elementary reason that the principal species, *Astrocaryum*, bears no edible product at this time. We noted the first use of *Astrocaryum* in late March when the fleshy exocarp of the fruits began to ripen. Shortly afterward, in early April, the trees dump their large fruit clusters *en masse* at the bases of the trunks, ending the brief interval when the sweet, mango-like exocarp is available to monkeys. The fruits then rot, leaving the nuts which lie in heaps on the ground for another month before they begin to attract attention.

During the June sample, *Astrocaryum* nuts accounted for more feeding time than any other item in the diets of both *Cebus* species. However, there were major differences in how the two species used the nuts and in their ease of handling them. *Cebus apella* ate nuts in lieu of fruit; the total feeding time for the sample was about the same as in other samples. In *albifrons* the time spent on nuts was in addition to a normal amount of time in fruit trees; the total feeding time was two-and-a-half to four times longer than in other samples. Moreover, *albifrons* spent considerably more time feeding on the nuts (130 min. per monkey-day) than did *apella* (45 min. per monkey-day). The reason for this was the vastly greater handling time for *albifrons*. A sound nut was too hard to be cracked by even the largest member of the troop, as was attested by numerous unsuccessful attempts. Opening one required strategy and persistence. Herein lies a curious tale of dependency of monkey on insect.

In keeping with Janzen's (1970) observations on seed predation in tropical forests, most of the nuts that had fallen around the base of parent trees had been infected by a bruchid beetle (Kiltie 1980). Damaged nuts were easily distinguished by being lighter than intact nuts, by having a small hole through which the insect had exited, and, evidently, by smell. There was great variability in the extent of damage. Some were completely unscathed (a small minority), some were partially eaten, and others (the majority) were completely hollow or rotten. From the point of view of the *albifrons*, the intact nuts were of no interest because they were too hard to open, and the hollow ones, though easily cracked, offered no reward. The challenge then was to find nuts that had been partially consumed by a beetle larva. A portion of the endosperm inside would still be undamaged, and the resistance of the shell would be reduced by the presence of a cavity inside. As such optimal nuts were a small minority, successful foraging depended on being selective and on exercising keen powers of discrimination.

During the June sample the *albifrons* troops spent long hours on the ground under groves of *Astrocaryum* palms. The animals were extremely selective, often checking and rejecting twenty or more nuts before finding one that was suitable. The preliminary screening was conducted by smell. (Presumably byproducts of the bruchid larva and decayed endosperm both emit characteristic odors.) Nuts that passed the smell test were then tapped against a branch or another nut, or bitten to assess their resistance. Many were rejected at this stage, as well as earlier when first sniffed. When an animal found one that seemed to possess the desired qualities, it usually went up onto a low branch for the opening operation. There it would often bash the nut vigorously against the branch, and then begin to bite it with its premolars. After a bite or two the nut would be rotated to a slightly different position and bitten again. If it failed to yield, it would be rejected, and the whole selection process would begin anew. If the nut cracked, the endosperm would be laboriously excavated from the shell, using canines or fingernails to pick at the firm material.

Not all individuals followed the same procedure. Some seemed to be more selective than others, and some preferred biting to tapping or vice versa. The most distinctive behavior was that of a low-ranking adult female who consistently opened nuts not by biting but by knocking two of them forcibly together. Holding one in each hand, she would bash them together two or three times. Then, holding one firmly as a mallet, and rotating the other, she would bash them again until usually the rotated one cracked. Struhsaker and Leland (1977) described this behavior in *Cebus apella*, also in connection with opening *Astrocaryum* nuts.

Here the comparison between *apella* and *albifrons* is especially revealing. Handling times for *albifrons*, including selection, averaged 5–10 minutes for each nut successfully opened. And then, we must remember, the reward was a half-eaten endosperm. By virtue of their larger size and stronger jaw musculature (Kinzey 1974), *apella* had no such difficulty. They routinely crushed intact nuts with a single bite. Half-grown juveniles seemed no less capable of this than adults. The foraging habits of *apella* were different as well. Instead of searching for fallen nuts under the trees, where they sometimes lay in heaps, the *apella* ferreted them out, one by one, from the tangle of spines, debris, and dead petioles that rim the crowns of the trees themselves. Although we were unable to confirm it directly, the *apella* appeared to be finding intact nuts. Possibly the few scattered ones that remained hidden in the crowns had not been discovered by bruchids. If so, this would explain the preferential foraging of *apella* in the crowns, even though nuts were more plentiful on the ground.

THE IMPORTANCE OF PALMS AND FIGS TO *Cebus* AND *Saimiri*

We have just seen how *Cebus* use palms. How much they use them is indicated in Table 5.8. They form a comparatively minor component of the diet except during the critical period of fruit scarcity in the early dry season when they assume a preeminent importance. A similar pattern holds for figs, except that the period of intensive use extends into the early part of the rainy season. Palms and figs together account for 73 and 97% of the feeding time of *Cebus apella* and *albifrons*, respectively, through the entire five-month dry season.

Saimiri is almost entirely excluded from palms by its small size. The time recorded for the May sample represents an extraordinary situation in which the monkeys scavenge *Scheelea* nuts that have previously been opened and partially eaten by *Cebus*. By itself *Saimiri* is unable to exploit this resource,

Table 5.8
Importance of palm nuts, figs and all other plant materials in seasonal samples (percentage feeding time).

	Season											
	Early rain (Oct.–Dec.)			*Late rain (Jan.–Apr.)*			*Early dry (May–July)*			*Late dry (Aug.–Sep*		
Species	*P*	*F*	*O*	*P*	*F*	*O*	*P*	*F*	*O*	*P*	*F*	
Cebus apella	1	17	82	4	6	90	64	9	27	14	49	3
Cebus albifrons	<1	17	82	4	19	77	56	41	3	10	73	1
Saimiri sciureus	–	59	41	–	7	93	15	77	8	–	56	4
Saguinus imperator	–	62	38	<1	–	>99	–	–	100	–	8	9
Saguinus fuscicollis	–	14	86	<1	4	96	<1	–	>99	–	–	

but by associating with *Cebus* it can in a limited way. The major dry-season resource for *Saimiri*, however, is figs. In the May sample they accounted for over 90% of the feeding time on soft fruit and 77% of all feeding time. Figs are important at other times of the year as well, accounting for over 50% of the feeding time in all but the late rainy season sample. *Cebus albifrons* is also a very heavy fig eater but is not dependent on figs to the same degree as *Saimiri* because of its facultative ability to eat *Astrocaryum* nuts.

Specializing on figs to the extent done by *Saimiri* and *Cebus albifrons* requires certain definite behavioral adaptations. The availability of figs in space and time is extremely irregular. The trees tend to be huge and produce enormous crops, but on no particular schedule. A given tree may fruit more than once a year, or it may fruit in a different month on each of a succession of years (Janzen 1979). Moreover, the species tend to be rare. Figure 5.1 shows the locations and sizes of known individual fig trees in the study area. Although a number of species are present, the density of most of them is in the range of one to five individuals per km². It is thus obvious that any animal which is dependent on a steady supply of figs must be highly mobile and must have a very large home range. *Saimiri* and *Cebus albifrons* conform to

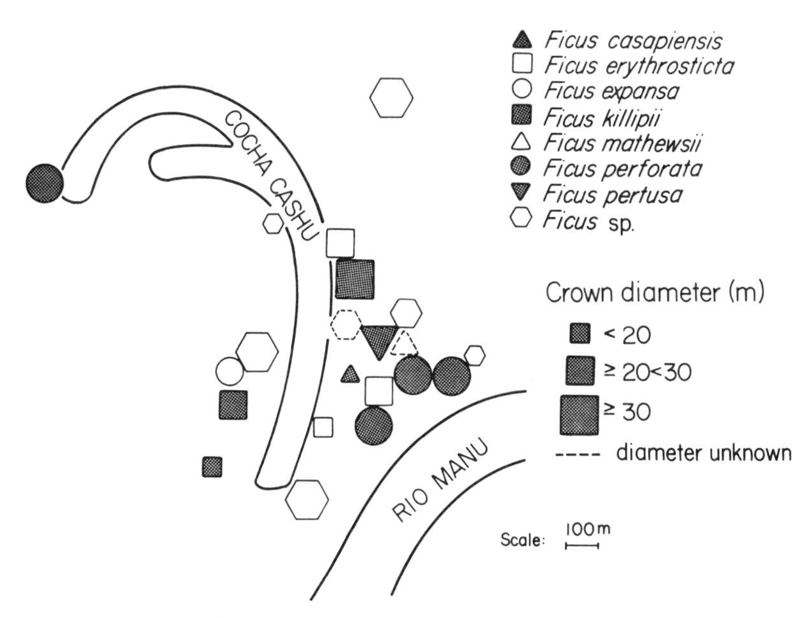

Figure 5.1 Size and location of fig trees (*Ficus* spp.) within the study area. *Ficus insipida*, a common species in riparian successional vegetation, and *Ficus trigona*, abundant in swamps, are not illustrated. Neither is known to fruit at Cocha Cashu during the critical period of the dry season.

these expectations in full measure. With home ranges of more than 250 ha and about 150 ha, respectively, they cover a far larger area than most arboreal primates (cf. Chapter 8). If either were to confine its activities to a smaller area, it would face the virtually certain prospect of being without food for long periods each dry season. *Cebus apella* survives within its much more limited home range by virtue of its extraordinary ability to exploit palms.

The Use of Major Resources by *Saguinus*

We turn now to the tamarins and the distinctive properties of their food supply. It should be recalled that the two species live in permanent mixed associations in which family groups of each species join to make up the territorial unit. The plant resources harvested by the two species are virtually identical. Within each territory a single set of trees is shared by the two resident groups. At any moment in time, a large number of trees are in current use, even though only a small fraction of them are visited each day. Return visits to individual trees are conducted every few days over periods of weeks or even months.

This practice exposes an important difference between the resources used by *Saguinus* and the resources used by most of the other primates in the community. Practically all of the plants that are heavily exploited by *Saguinus* ripen their crops in a piecemeal fashion. Although we were months into the study before we recognized this, the fact is of the utmost significance for the ecology and behavior of *Saguinus*. A prolonged ripening schedule implies two things about the availability of the fruit: (1) that only a small number is ripe and ready for harvest on any occasion, and (2) that a reliable, though scanty, supply can be harvested on each visit over many weeks. Fluctuations in resource levels are thereby strongly damped, and the loci at which resources can be obtained are fixed in space for extended periods. The rate of harvest that can be realized from any given plant is limited by the ripening rate of the crop.

This stands in strong contrast with most of the fruit species used by the larger primates. The crops tend to ripen within a narrow time span, both within individual crowns and over the whole population of a species (figs excepted). Fruit that is not harvested may fall off and become unavailable, at least to arboreal frugivores. Concentrated crops permit the high feeding rates that are important to animals of large body size. In most of the plants used by *Saguinus*, feeding rates are low. To collect the scattered ripe fruits, a tamarin must move over the entire crown, picking just a few on each branch. The yield per unit time is not sufficient to satisfy a *Cebus* or *Saimiri*, especially if alternative resources are available. Varied ripening schedules thus offer plant species an important means of discriminating between potential fruit dispersers.

One plant that is used intensively by *Saguinus* over an extraordinarily long period is *Celtis iguanea* (Ulmaceae). Both *Cebus apella* and *Saimiri* eat *Celtis* fruits, but only casually as they happen to chance upon the vines in the course of their foraging. This is indicated by the huge disparities in use levels between these species on the one hand, and the two *Saguinus* and *Callicebus moloch* on the other (Table 5.9). From the point of view of the latter three species, *Celtis* is the predominant resource during April, May, and probably June, and for *Saguinus* it continues to serve as a minor resource into August. *Cebus apella* and *Saimiri* use *Celtis* in a much more limited way, but the troops are much larger than those of *Saguinus*. Correcting for this (Table 5.9), we see that the difference still holds. The average *Celtis* vine receives twice as much use from either *Saguinus* as it does from *Saimiri*, and the disparity with *Cebus apella* is far greater. *Callicebus moloch* is the closest competitor of *Saguinus*, at least during the *Celtis* season. The main conclusion to be drawn here is that slow-ripening crops are essentially unavailable to the larger primates, not for lack of productivity but because the fruits cannot be harvested fast enough and in sufficient quantity from any one plant to satisfy the demands of a large animal, much less one that lives in sizable troops. *Saguinus* (and perhaps

Table 5.9
Use of *Celtis iguanea* fruits by five primate species in successive sample periods.

Species	*Dates of sample*	*Rank of importance of Celtis in diet during sample*	*Use of fruits from point of view of:*	
			Plant (monkey-min/day)	*Monkey (min/monkey-day)*
Cebus apella	April 2–8	10	1.1	0.1
Saguinus imperator	April 9–22	1	430	86
Saguinus fuscicollis	April 22–May 1	1	350	78
Callicebus moloch	May 2–7	1	190	63
Saimiri sciureus	May 8–May 30	3	180	5.1
Cebus apella	June 12–July 1	8	13	1.1

Callicebus) subsist on such crops. They are able to do so by virtue of their relatively low metabolic demands and small group sizes.

Celtis iguanea is one example of what I shall call a major resource of *Saguinus*. For two or three months it provides the main carbohydrate supply, accounting for up to 90% of all feeding time (Tables 5.10 and 5.11). The tendency for the tamarins to concentrate their feeding activity on one plant species at a time is remarkable. It is a habit they appear to follow at all times of year, regardless of how many alternative resources may be available within their territories. This was most clear in the November *fuscicollis* sample. An ample variety of fruit was available, as judged from the fruit trap data, and from the observations that *Cebus apella* used 31 fruit species in the preceding sample, that *C. albifrons* used 29 in the sample immediately following, and that the *fuscicollis* themselves used 14. Nevertheless only one species, *Guatteria* sp. (Annonaceae), was responsible for 61% of the total feeding time. *Cebus* and *Saimiri* may show this degree of concentration during the dry season when alternative resources are scarce, but at other times their feeding is spread over many species. Yet for unknown reasons the tamarins single-mindedly dedicate themselves to harvesting one plant species at a time, while largely ignoring many others that are avidly consumed by the larger primates in the community. It is not that all these fruits are too large or otherwise unmanageable for the tamarins; they do in fact sample a good many of them. They just do not eat them in any quantity.

Table 5.10
List of major resources used by *Saguinus* spp. at Cocha Cashu, and their salient characteristics.

Species	*Type of plant*	*Harvested product*	*Height of product (m)*	*Period available*
Celtis iguanea	Liane	1-cm drupes	12–45	March–August
Combretum assimile	Liane	Nectar	20–45	July–early August
Quararibea cordata	Tree	Nectar	30–45	Late July–August
Pourouma cecropiafolia	Tree	1-cm drupes	20–30	August–November
Guatteria sp.	Tree	1-cm drupes	25–35	October–December

NOTE: The list is complete for the period of late March to early December; we have no observations for the interval between December and March.

Table 5.11

Use of major resource plant species by *Saguinus imperator*, *S. fuscicollis*, and *Callicebus moloch*.

Species	*Sample*	*No. in troop*[a]	*No. contact days*[b]	*Total No. trees used*	*No. of trees belonging to major resource species*	*Total tree species used*	*Daily feeding time (min./monkey-day)*	*Percentage feeding time on major resource species*[c]	
Saguinus imperator	August 1975	3	8.5	⩾108	43	19	157	Q + Ce	51
	October 1976	4	3.2	12	4	6	46	F	62
	April 1977	5	11.0	105	74	16	95	Ce	90
	July 1977	4	7.2	64	50	7	167	Co + Ce	92
Saguinus fuscicollis	November 1976	3	6.8	43	16	14	57	G	61
	April 1977	4.5	9.5	78	59	12	89	Ce	88
	July 1977	3	7.8	74	57	14	142	Co + Ce	88
Callicebus moloch	May 1977	3	4.9	37	27	5	98	Ce	66

[a] Number of independently locomoting individuals.

[b] Total number of contact hours for the sample divided by the mean length of the daily activity period.

[c] Key to major resource species: Ce = *Celtis iguanea*; Co = *Combretum assimile*; F = *Ficus* spp.; G = *Guatteria* sp.; Q = *Quararibea cordata*.

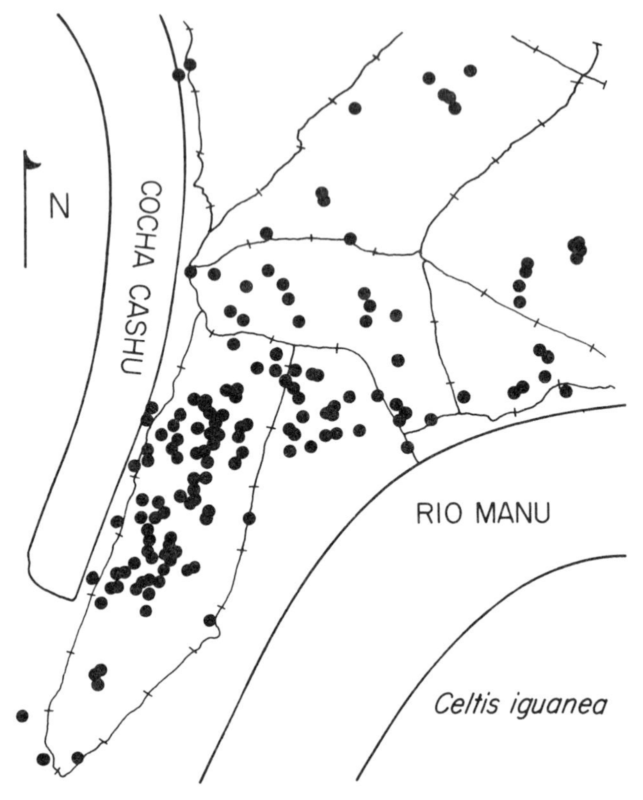

Figure 5.2 (Left) Known locations of *Celtis iguanea* and (right) *Combretum assimile* vines, two major resource species for *Saguinus*.

In the course of a year the *Saguinus* at Cocha Cashu rely on a minimum of five and a maximum of probably no more than six or seven major resource species to supply the bulk of their carbohydrate and caloric needs (Table 5.10). All of these species share a suite of common characteristics. They are relatively common plants, having densities of two to many individuals per hectare within occupied *Saguinus* habitat (Fig. 5.2, and compare against fig map, Fig. 5.1). Their fruits or flowers mature in a piecemeal fashion, assuring a prolonged period of availability. All are trees of medium size or lianes that bear their flowers and fruits in the canopy, though not at the highest level. Collectively, the five major resource species that are exploited between March and December provide a continuous and relatively stable food supply for *Saguinus* throughout the period when fruit is generally scarce in the forest. Continuity of the resource base is assured by the fact that the production period of each species broadly overlaps those of the next one or two species

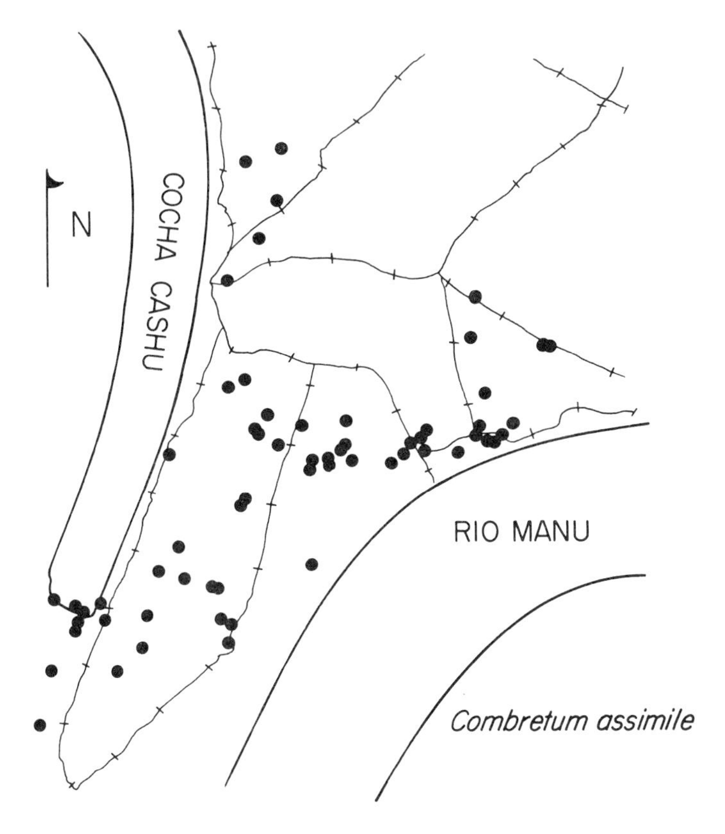

in the sequence. Although we do not know what the tamarins eat between December and March, we can safely assume that suitable fruits are available because this is the period of peak production by the forest as a whole.

Feeding time for *Saguinus* varies with the nature of the resource being exploited (Table 5.11). Drupes of *Celtis* can be harvested and eaten in an intermediate amount of time, measured at 89 to 98 minutes per monkey-day in three independent samples. *Celtis* fruits are split open in the mouth. The tough and slightly rough exocarp is then ejected as the animal chews on a sweet mucilaginous coating around the large central seed. Sometimes the seed is ejected, but more often it is swallowed. *Saguinus* feces in April consist of little more than polished *Celtis* seeds. *Guatteria* fruits are handled somewhat differently. The outer covering is membranous and soft. It adheres to a cortical layer of fleshy pulp that surrounds the spindle-shaped seed. The animals pick individually selected ripe fruits with their mouths (Cebids use their hands to pick fruit), chew off the pulp, and eject both the seed and the exocarp. It can be surmised that the caloric yield per fruit is greater or the handling time less than for *Celtis*, because the feeding time was markedly lower in November (57 min. per monkey-day) when *Guatteria* was the major resource. In July the major resource is the nectar of *Combretum assimile* flowers. These are carried on stout-pedicelled horizontally arrayed spikes. Each spike contains from 80 to 110 individual flowers which are held in an upright position. The flowers on a given spike open synchronously, or nearly so, and produce nectar for two days after opening. Although the amount of nectar produced per flower is highly variable within a spike, the variability between spikes is much less. If harvested early in the morning prior to any visitation, the average spike yields 2 to 5 ml of dilute nectar. It is doubtful that tamarins can extract nectar as efficiently as a human being armed with a hypodermic syringe. One to 3 ml per spike would be a reasonable guess as to what the animals can do. At that low yield it is obvious that a great many spikes would have to be harvested to provide a meal. This is reflected in the feeding times recorded during the July samples (142 and 167 min. per monkey-day) when *Combretum* flowers constituted the principal resource. A similar result (157 min. per monkey-day) was obtained in the August 1975 *imperator* sample when the nectar of *Quararibea cordata* was the leading resource. The greatly expanded feeding times suggest that nectar is a poor-quality resource relative to the fruits that are exploited at other times of year.

Minimum feeding times were recorded during the brief October *imperator* sample. This was the only occasion on which figs featured prominently (62%) in the diet of *Saguinus*. Relative to drupes and flowers, figs yield food at a high rate because the crops ripen synchronously and can be harvested non-selectively, and because the entire fruit is edible and does not require careful manipulation within the mouth. The ease of handling is reflected in the very low feeding time (46 min. per monkey-day) recorded for this sample.

Summary and Conclusions

Soft fruit is the principal plant resource consumed by all five species and makes up nearly 100% of the vegetable component of their diets during the period of abundance. When the fruit supply dwindles at the end of the rainy season, the animals turn to a wide range of additional plant products to supplement their diets (nuts, nectar, pith, meristem, etc.). The fact that most of these alternative products are available throughout the year, but are eaten only when fruit is in short supply, suggests that they are foods of last resort.

The scarcity of fruit during the dry season is reflected in a marked drop in the number of plant species used by the monkeys at that time. The reduction is particularly pronounced in *Cebus* and *Saimiri*, which are attracted to a wide variety of fruit species, as opportunity permits. In contrast, the diversity of plant species in *Saguinus* diets is much lower and more constant, due to their peculiar habit of concentrating single-mindedly on just one or two resources at a time.

At the taxonomic level there is much overlap in the fruit species used by the various monkeys. Genera that are prominent in the diet of one species are likely to be exploited by others as well. The rarity of many plant species limited the opportunity to record the full extent of their use by members of the primate community. Altogether we recorded the use of over 170 species of plants in 55 families.

A number of characteristics of fruits and fruit trees were examined to determine their roles in promoting the differential use of plant resources by monkeys. Consumption patterns seem to reflect more what was available than what were intrinsic preferences. The size of fruits had a surprisingly weak influence on the species exploiting them. The large *Cebus* species often fed heavily on fruits of the smallest-size class (<5 mm, mostly *Ficus* spp.), while the small *Saguinus* used large fruits (>4 cm) no less than did the other species. Hardness is a quality that effects more differential use. This is particularly true in the case of palm nuts, which only the *Cebus* were able to exploit. Among the fruits eaten by monkeys, bright yellow and orange coloration predominates. It is surmised that these colors produce the greatest possible contrast against a background of dark green foliage.

The heights of trees used for feeding do not differ greatly among the species. Little fruit is harvested from the understory even though a good deal of insect foraging is done there. Most fruit eating is concentrated in the middle strata of the forest (in 11–30 m trees) except in *Saimiri*, which logged half its feeding time in trees that were over 30 m tall. In contrast, *Saguinus* only rarely ascends to such heights. Crown diameter proved to be much more important than tree height in discriminating the activities of the monkeys. *Saimiri* characteristically fed in the largest trees (modal class >30 m), followed by *Cebus albifrons* (modal class 21–30 m). This resulted from the strong preference for figs shown by both species. *Cebus apella*, which uses palms far more and figs less, fed mostly in small trees (modal class 6–10 m). This was true of the *Saguinus* as well, but with the striking additional proviso that they completely shunned large trees.

Although *Cebus* spp. and *Saimiri* eat very similar assortments of fruit in the rainy season, their feeding behavior undergoes a marked divergence in the dry season. *Cebus apella* finds little soft fruit within its comparatively small home range, but makes up the deficiency by consuming large quantities of palm nuts supplemented with pith from various well-protected plant organs. Nuts, especially of the *Astrocaryum* palm, are an abundant resource at an average density of 40 trees per ha. Even juvenile *Cebus apella* break them with apparently little difficulty. But this is not true of *Cebus albifrons* which require a 5–10-minute handling time for each nut. Although palm nuts are used very extensively by *albifrons*, the preferred resource is figs. Because fig trees are widely scattered and bear fruit at unpredictable times, the animals must travel long distances to locate ripe crops. It is this fact that accounts for the large home ranges of *Cebus albifrons* and *Saimiri*.

The differences between the species can be summarized quite simply (Table 5.12). *Cebus apella* survives the dry season by concentrating on palms which offer a reliable and abundant alternative to fruit. *Cebus albifrons* and *Saimiri* travel long distances to find figs which, however, are not a reliable resource. To get themselves through the inevitable figless interludes, the two species employ divergent tactics. The *albifrons* resort to eating *Astrocaryum* nuts, which they do slowly and with difficulty, while the *Saimiri* concentrate exclusively on insect foraging at a rate of return that may be below the caloric break-even point. While these last-resort measures are probably adequate in most years to assure a high rate of survival until the return of better times, the nutritional status of the populations may suffer temporary declines.

The situation of the tamarins is strikingly different. Their territories appear to contain an adequate supply of resources at all times of year. This results from the fact that their principal food plants possess a highly distinctive set of characteristics. The individual plants are of small to medium size and the species are common within the defended territories. But most importantly, the plants have exceptionally long bearing periods, ripening their crops piece-

Table 5.12
Differential use of food resources during the early dry season (May–July).

Species	*Early dry season feeding habits*
Cebus apella	Heavy use of *Scheelea* and intact *Astrocaryum* nuts, supplemented with pith, meristematic tissue of palms, and figs when available.
Cebus albifrons	Concentrated feasting on figs when available, alternating with laborious exploitation of bruchid-infested *Astrocaryum* nuts.
Saimiri sciureus	Concentrated feasting on figs when available, alternating with periods of pure insectivory.
Saguinus imperator and *S. fuscicollis*	Exploitation of a sequence of resources that are too diffuse to be of major significance to larger monkeys: *Celtis iguanea* (April–August), *Combretum assimile* (July, early August), *Quararibea cordata* (August).

meal over many weeks. These characteristics not only assure great stability in the food supply, but assure as well that most of the supply is sequestered from the depredations of larger monkeys (*Callicebus* excepted). The reason for this is that tamarin resources occur in many scattered pockets, each one of which yields so little per visit that a larger animal does not find sufficient reward.

This observation brings us to the major result of the chapter. The most important characteristic of fruit from the point of view of differential exploitation by primates is not its size, texture, color, construction, or taxonomic status; it is its characteristic degree of concentration in space and time. Most monkeys can eat most species of fruit. But large-bodied monkeys that travel in sizable troops cannot profitably or efficiently exploit a resource that is diffused in tiny packets over a large area. Instead, they must seek out scattered but concentrated sources which, when reached, allow high feeding rates. In comparing the primates at Cocha Cashu, it is clear that they fall along a spectrum of behaviors with regard to dependence on concentrated resources. *Saimiri*, with its absolute requirement for figs in the dry season, is at one extreme. *Cebus albifrons* follows next in that it prefers concentrated resources but can make due with a diffuse resource (palm nuts) when necessary. *Cebus apella* is far more proficient at using palms than *albifrons*; hence it can live in a smaller area but nevertheless shows signs of preferring fruit, as we shall see in Chapter 8. And finally, at the far end of the adaptational spectrum, *Saguinus* makes its living on resources that come in tiny, scattered, incremental units. Widely varying levels of concentration of fruit resources thus provide the basis for a primarily behavioral mode of adaptation by which the fruit supplies of the forest are allocated differentially to the various members of the primate community. The other major mode of adaptation is morphological, mainly expressed through differences in body size. Important consequences of size were shown in this chapter in the force required by *Cebus apella* to crush *Astrocaryum* nuts, and in the ability of *Saguinus* to survive on resources that fall beneath the notice of larger species. These two themes—the behavioral and morphological adaptations for feeding—will occupy us for most of the rest of the book.

6 Foraging for Prey

The importance of animal prey to the five species was suggested in the analysis of their time budgets. Foraging was for all a major, if not the major, activity, occupying from 15 to 49% of their waking lives. Such substantial investments of time imply that the capture of prey is a matter of necessity and not merely a casual pastime. We may thus suspect that differences in foraging behavior may play a substantive role in defining ecological relationships among the species.

Before we plunge into the details of how each of the species hunts for prey, it will be instructive to review the ways in which prey, as a resource, differs from fruit and consider the implications of these differences for possible mechanisms of resource partitioning. Relative to fruit, prey items (and here we shall be speaking primarily about insects and other arthropods) are widely diffused in the environment. There are myriads of species, each with its own characteristic phenology, lurking places, and devices for self-preservation, and few, if any, offer enough biomass to permit the specialization of a comparatively enormous predator such as a monkey. Monkeys thus have little choice other than to be generalists, at least at the level of prey taxon.

As predators, monkeys are searchers rather than pursuers (MacArthur and Pianka 1966); that is, most of their hunting time and effort is devoted to locating prey which, once found, are relatively quickly captured and eaten. This can be contrasted with the behavior of a pursuit predator, such as a cheetah or falcon, which invests concentrated bursts of effort in pursuing and killing relatively large prey. While pursuit predators are predicted by optimal foraging theory to be quite selective, searchers are expected to be unselective and to take nearly every prey item they find, provided that it is inherently palatable. More specifically, any item should be included in the diet so long as the ratio of its food value to the time needed to capture and ingest it (handling time) is greater than the net rate of energy intake without it (Charnov 1976; Pyke et al. 1977). For searchers, capture and ingestion times are generally low, so that all but the smallest prey items should be eaten as found. Of course there will be exceptions, such as prey that are too large to be managed, or ones that possess armored exteriors, noxious chemicals, or intimidating defenses such as stings. Later, we shall find that selectivity increases when the capture of small prey reduces the probability of capturing large ones, but in general we can anticipate that monkeys will be nonselective in their choice of prey.

Even if nonselectivity were practiced by the various species, it does not follow that all animals should converge on the same search routine or take the same distribution of prey taxa. Instead, it stands to reason that each predator should follow the hunting routine that brings in the greatest rate of return. Through its own peculiar morphological and locomotory capabilities, any species will be better suited for some tasks than for others, and consequently will be constrained by its successes and failures to forage in certain ways. Any given pattern of behavior will lead to definite consequences in an animal's hunting by influencing the rate at which prey are discovered and the types of prey that are encountered. Here it becomes important to distinguish between preference and opportunity. Other things being equal, preference can be expected to increase in proportion to prey size. Yet it is readily possible that two predators with identical preferences might capture prey of distinctly different average size as a result of the determinate consequences of differing search and capture techniques. Later in the chapter we shall see how the conflict between preference and opportunity can sometimes lead to paradoxical outcomes.

Another kind of trade-off that enters into the determination of foraging behavior is found in the balance of techniques employed. If an animal is capable of performing several tasks with reasonable skill, the relative rates of success it achieves with the various tasks will regulate the proportions with which they are employed. The relative success rates will depend on the distribution of prey in the environment, and this, in turn, will depend not only on the intrinsic abundance and behavior of the prey, but on the presence of competitors. The perceived effect of competitors will be to reduce the rates at which prey are encountered using any given search technique. To the extent that competitors are selective in their hunting methods, the relative rates at which prey are captured by alternative techniques will be perceived to vary, and the animal can consequently be expected to emphasize the most effective ones in its repertoire. This is a mechanism that can promote the specialization of behavior, either in ecological or in evolutionary time. However, as competitors were always present during our study, we shall not be able to demonstrate their effects directly.

Other predators are not the only factors that may alter the distribution of prey in the environment. Arthropod communities may undergo drastic changes in their abundance and composition in tropical forests in concert with the annual cycle of seasons (Janzen 1973; Ricklefs 1975; Smythe 1974). Such changes can be expected to motivate appropriate shifts in tactics on the part of predators, within the limitations of their capabilities.

The primary intent of this chapter is to show how each species' foraging behavior is both constrained and guided by its morphology (principally size) and locomotory pattern, and how the distinctions between the species lead to differences in where and how prey are captured and in what kinds of prey are taken. Once I have described our method of taking foraging data, the rest of the chapter shall be devoted to presenting details on how the five species find and capture their prey. To provide a full description of how an animal forages, one needs several kinds of information: the substrates searched, the types of manipulation involved in searching, the characteristics of perches used while searching, the repertoire of capture techniques, and the taxa of prey captured. After presenting data on each of these for the five species, we shall examine more closely the results for *Cebus apella* to show that there is appreciable intraspecific variation in behavior as manifested by different age and sex classes of individuals. We then turn to a consideration of the temporal aspects of foraging, with a discussion of both diurnal and seasonal patterns of variation. The final discussion takes up the significance of body size as it relates to the modes of resource partitioning in omnivorous primates and to its antithetical effects in the evolution of improved ability to compete via interference or via exploitation.

Method

All data on foraging behavior were taken in timed sequences. When an animal came into view in a place affording good visibility, the observer punched a stopwatch and began accumulating a mental tally of everything it did. If the subject then passed out of view before 30 seconds had elapsed, the sample was discarded. All sequences lasting more than 30 seconds were recorded, but any that continued as long as 5 minutes (few did) were terminated at that point. This was to avoid overrepresentation of stationary activities such as the bark stripping sometimes engaged in by *Cebus*.

In the sequences we recorded the age and sex (or individual identity when known) of the subject animal, the supporting structures (tree, vine, palm, etc.), each substrate searched (branch tip, live leaf, dead leaf, etc.), the accompanying search movement (open rolled leaf, pull branch to, bit open hollow twig, etc.), the success of each search (+ or −, as judged by whether the animal chewed), the motion used to capture each prey item (snatch, grab, lick leaf, etc.), and, whenever possible, the type of prey captured and its length (1-cm caterpillar, 3-cm katydid, etc.). All this was recorded at the end of the sequence in a conventionalized shorthand notation. Sometimes the distance the animal moved during the sequence was also included. Although remembering all the details of a sequence was difficult at first, our retention improved with practice. We decided against tape recorders at the outset because using them would have meant an additional step in the processing of data, which was already an enormously time-consuming operation.

Comparative Foraging Behavior

Qualitative Sketches

Primates are remarkably versatile and opportunistic in their foraging behavior, engaging in many kinds of search and capture activities. Birds tend to be far more stereotyped in their movements and for this reason it is routine in the ornithological literature to characterize species as gleaners, probers, hawkers, etc. (Fitzpatrick 1980). The foraging of primates does not lend itself to such facile characterization. There are major differences between the species we studied, but these differences are more quantitative than qualitative, and pertain not only to capture techniques (usually emphasized in the ornithological literature) but also to the substrates searched and to the behaviors employed in searching. Since our analysis must necessarily focus on the components of foraging behavior one at a time, there is a risk that the important differences in the activities of the five species will be blurred by the welter of quantitative detail. It will therefore be advantageous to begin with a series of thumbnail sketches that will convey a rough mental image of how each species goes about obtaining prey.

Cebus apella, being the most heavyset and powerful of the species, engages in a great deal of what can be termed destructive foraging. It bites open bamboo canes, hollow dead twigs, and dead palm rachides; it is particularly attracted to palm crowns, where it rummages through the debris that accumulates in the funnel-like apical regions (Izawa 1979). It often rummages through matted vine tangles, sending down showers of dead stems and leaves. Another common pursuit is the stripping of bark from dead trunks and limbs. All these activities require strength. The object is to expose hidden prey that cannot escape by flight, such as beetle larvae and the brood of ant colonies. *C. apella* does do a fair amount of leaf and branch foraging while in transit, but prolonged foraging stops are nearly always dedicated to the kinds of destructive searching described above.

In many ways *C. albifrons* is intermediate between *C. apella* and *Saimiri*, and this applies to its foraging behavior as well. It is more cursory than *apella*, devoting more time to superficial searching of foliage and less to destructive branch breaking. It spends less time in palms, and when in palms it concentrates more on the fronds and less on the debris in their centers.

One notable idiosyncrasy of *albifrons* is its fondness for hymenoptera. Even large paper nests teeming with formidable wasps are not immune to attack. Having discovered such a prize, an *albifrons* will usually consider it carefully, approaching from one angle, then backing off to check another, stopping in between to stare intently at its target. Finally, it will rush in, either to bat the nest to the ground or to snatch it from its attachment. If it takes the nest, it tucks it under one arm and then flees at top speed across several tree crowns. The wasps may attack in force, but the fur of *albifrons* is long and fluffy and protects all but its hands and face. Once the monkey stops, it frantically brushes off the wasps that are tangled in its fur, just as you or I would. Such raids are not without their risks, for the animals wince conspicuously when stung. Indeed, the risks may sometimes be perceived as too high, for we have seen instances in which an *albifrons* has retreated from a particularly large nest after having given it close scrutiny for some minutes. Perhaps because of the greater protection afforded by its long fur, *albifrons* takes wasp nests far more often than *apella*.

In progressing down the size scale from *apella* to *albifrons* to *Saimiri*, there is a pronounced decrease in destructive foraging and a complementary increase in superficial gleaning. The members of a *Saimiri* troop move like so many automatons, progressing in almost continual motion. Individuals pause only momentarily to open a dead leaf, to turn over a leafy branch, or to lick a pupating insect from a twig. They are on the go nearly constantly, and even when a group as a whole is stationary, the monkeys persist in their searching, meandering here and there without any general direction. Though by no means as stereotyped in its behavior as many insectivorous birds, *Saimiri* is nevertheless decidedly less varied in its foraging than the two *Cebus* species.

The same can be said of the *Saguinus* species which pursue tactics that differ radically from those of the larger monkeys. In its preference for leafy substrates, *S. imperator* closely resembles *Saimiri*, but its search behavior is distinct. Instead of manipulating substrates to see whether they carry attached or concealed insects, *imperator* more often scans while nearly motionless on a perch. While a *Saimiri* always focuses on what is in its hands, an *imperator* looks for prey at much greater distances. Where a *Saimiri* often grabs an item with one hand or pulls a branch to bring an attached egg case up to its mouth, an *imperator* frequently pounces, trapping the prey between two cupped hands. The difference is that most of the prey of *Saimiri* are small and sluggish or immobile, while many of those of *imperator* are large and capable of rapid escape. *Saimiri* hunts by dogged persistence; *imperator* by stealth and ambush. In this the two could hardly be more different.

We are now left with *S. fuscicollis*, which is even more distinctive in its habits. In keeping with its propensity for cling-and-leap locomotion and its great agility on trunks, *fuscicollis* is a bark and cavity specialist. The animals do nearly all their foraging on vertical surfaces, mainly large fluted or buttressed trunks or heavy liane stems. They are obsessively curious about cracks and crevices and unhesitatingly reach into any dark knothole they discover (Yoneda 1981). Hollow trunks and limbs are thoroughly investigated, and one often hears reverberant thumps and scratches as the animals scamper about inside. In this behavior *fuscicollis* is absolutely unique among the primates at Cocha Cashu.

QUANTITATIVE INTERSPECIFIC COMPARISONS

Substrates searched. There are major statistical differences between most of the species in their use of foraging substrates (Fig. 6.1). The *Cebus* species and *Saguinus fuscicollis* hunt a good deal on trunks and branches; but as described earlier, the *Cebus* are destructive searchers, either biting open hollow twigs or stripping dead bark, while *S. fuscicollis* is cursory and superficial. *C. apella* searches more in palms (32%) than any other species, while *albifrons* is second in this category (21%). *Saimiri* and *Saguinus imperator* are dedicated leaf foragers, devoting between 85 and 90% of their attention to this one class of substrate. *Saguinus fuscicollis* stands apart from all the others in avidly investigating knotholes, its principal foraging activity (62%).

Interspecific comparisons of overlap in substrate use indicate significant differences between all pairs of species except *Cebus apella-C. albifrons* and *Saimiri-Saguinus imperator* (Table 6.1). These pairs differ in other aspects of their foraging, as has been indicated already in the qualitative descriptions. *Saguinus fuscicollis* shows the lowest overlap with all other species due to its peculiar single-mindedness for knotholes. Intraspecific overlaps based on the seasonal samples are reassuringly higher than interspecific overlaps except for *C. apella*, which, as we shall see, is the most variable in its behavior, both within and between seasons. Intraspecific variability will be treated in a later section.

Table 6.2 indicates the diversity of substrate use for the five species based on the combined data from all samples. *Cebus apella* is the most versatile, followed by *C. albifrons*. Not surprisingly, the two leaf foragers show a low diversity of substrate use. *Saguinus fuscicollis* is intermediate due to its habit of searching bark and occasional leaves in addition to holes.

Support use. There are contrasts among the species in the kinds of perches they choose while foraging (Table 6.3). To a degree this is a simple reflection of size differences. A heavy-bodied *Cebus* obviously may not be supported by slender twigs and vine stems that may afford stable underpinnings for a *Saguinus*. Such differences, which are not registered in the data on substrate use, serve to amplify the spatial separation of foraging activities among the species. This conclusion is brought out in the frequency of use of vines as supports. With the exception of *Saguinus fuscicollis* and its proclivity for trunks, the use of vines increases with decreasing body size. *Saguinus imperator* does more than half its foraging in them. We now see a rather pronounced statistical separation of this species from *Saimiri*, something that was not apparent in the data examined previously. Palms constituted the modal

Cebus apella

N = 5,408

	Trees, vines	Palms	Other
Leaves	28	17	1.5
Branches	29	16	5
Knotholes	2	0.2	0.1

Other = 2.0%

Cebus albifrons

N = 3,635

	Trees, vines	Palms	Other
Leaves	47	14	2
Branches	23	6	3
Knotholes	1.6	0.1	0.8

Other = 2.6%

Saimiri sciureus

N = 4,637

	Trees, vines	Palms	Other
Leaves	83	4	1.1
Branches	7	0.5	0.1
Knotholes	0.1	0.0	0.0

Other = 4.2%

Saguinus imperator

N = 682

	Trees, vines	Palms	Other
Leaves	86	1.4	0.4
Branches	8	0.2	1.0
Knotholes	0.6	0.0	0.2

Other = 2.2%

Saguinus fuscicollis

N = 509

	Trees, vines	Palms	Other
Leaves	8	0.7	0.1
Branches	22	4	0.9
Knotholes	57	3	2

Other = 1.8%

Figure 6.1. Substrates foraged (percent of searches). Any manual inspection or manipulation of a substrate constituted a search. Data from all seasonal samples combined.

Table 6.1
Overlap in foraging substrates (based on data in Fig. 6.1).

Cebus albifrons	0.80			
Saimiri sciureus	0.43	0.62		
Saguinus imperator	0.42	0.61	0.94	
Saguinus fuscicollis	0.40	0.40	0.18	0.18
	Cebus apella	*Cebus albifrons*	*Saimiri sciureus*	*Saguinus imperator*

NOTE: Overlap index of Holmes and Pitelka 1968.

Table 6.2
Diversity of substrates searched in foraging.

Species	e^H
Cebus apella	5.0
Cebus albifrons	4.2
Saimiri sciureus	1.7
Saguinus imperator	1.7
Saguinus fuscicollis	3.2

NOTE: Expressed as the number of equally used substrates, $e^H = e^{-\Sigma p_i \ln p_i}$.

class for *Cebus apella*, which was not true of any other species. *Cebus albifrons* led in terrestrial foraging (10%).

Capture techniques. Here we consider just how the animals secured their prey. The types of behaviors employed reflect the fact that prey may be either exposed or hidden. If exposed (on a leaf, open bark surface, etc.), a prey item may be capable of rapid escape (by jumping or taking flight) or not (caterpillars, pupae, sluggish beetles, etc.). If hidden, prey can be sequestered in retreats that offer varying degrees of resistance to destruction. Each situation calls for an appropriate type of attack. Thus there is a tight correspondence between the potential mobility and state of exposure of the prey and the

Table 6.3
Supports used during foraging (percentage of searches).

Species	*Support*				
	Tree	*Vine*	*Palm*	*Ground*	*Other*
Cebus apella	34	21	**40.0**	0.5	3.7
Cebus albifrons	**47**	15	21.0	10.0	6.5
Saimiri sciureus	**61**	30	5.3	0.6	3.9
Saguinus imperator	39	**54**	2.6	0.2	3.6
Saguinus fuscicollis	**68**	18	12.0	1.2	1.1

NOTE: Modal class for each species in boldface.

capture method employed by the predator, a fact that is acknowledged in the organization of Table 6.4.

Exposed prey are the mainstay of the leaf foragers, *Saimiri* (70%) and *Saguinus imperator* (79%), which are light enough to move about easily on branch tips and vines. The two *Cebus*, on the other hand, are given to destructive foraging for hidden prey (76%), with *C. apella* engaging more frequently in the more strenuous classes of activity. Here we see for the first time a good separation of *C. albifrons* from both *C. apella* and *Saimiri*, as the modal values of all three species lie in different categories. *Saguinus fuscicollis* takes more hidden prey than any of the other species (89%), but obtains them by reaching into crevices and cavities rather than by biting or tearing away protective plant parts.

Taxonomic composition of prey. Interspecific differences in substrate use and attack mode must surely lead to differences in the types of prey captured (Hespenheide 1971). We attempted to identify prey whenever possible, but the vast majority of items were too small to be resolved or were hidden in the animal's hand while being transferred to its mouth. Only the larger items ($\geqslant$ 1 cm) could be identified with certainty. The available data are thus strongly biased with respect to the numbers of prey in each taxonomic category, though there is less bias with respect to overall caloric value, because the mass of a large prey item may be equivalent to that of several small items. We were not aware of any systematic bias between the species in our ability to identify large prey, once captured. More than anything else, prey identification depended on the proximity of the animal under observation. However, there were significant interspecific differences in the proportions of prey captured that were large enough to be identified. Many of the prey items eaten by

Prey capture methods (percentage of prey capture attempts, all samples).

Capture method	Snatch, pounce, "flycatch"	Grab, bring to mouth, bite leaf	Manual search of open or curled surface	Bite open, break with hands	Bite or break with full body force
Prey exposure	Exposed			Hidden	
	Visible and mobile	Visible and sluggish or immobile	Hidden in easily opened substrate	Hidden in moderately hard substrate	Hidden in very thick or hard substrate
Substrates	Leaf, open branch	Leaf, branch, palm frond	Leaf, bark, knothole, termite run, epiphyte	Dead branch, palm rachis, cane, bamboo, under bark	Large dead trunk, heavy bark, palm crown, termite nest
Cebus apella N = 5,157	2.8	21.1	31.8	**33.2**	11.1
Cebus albifrons N = 3,536	4.2	19.8	**43.7**	26.7	5.7
Saimiri sciureus N = 4,637	18.3	**51.4**	29.7	0.7	0.0
Saguinus imperator N = 691	29.8	**49.4**	19.7	1.2	0.0
Saguinus fuscicollis N = 509	7.2	4.6	**85.8**	3.0	0.0

NOTE: Modal class for each species in boldface.

Saguinus were large, while most of those eaten by the other species were small. We shall return to this difference in the discussion of capture rates.

Even though they contain unavoidable observational biases, the data on identified prey are nevertheless of interest because they reveal some strong contrasts that can be interpreted from the point of view of the different search and capture techniques of the species (Table 6.5). The two *Cebus* species, which as we have seen are very similar in their foraging habits, capture almost

Table 6.5
Taxonomic breakdown (%) of prey items large enough to be positively identified during capture.

Prey taxon	*Cebus apella*	*Cebus albifrons*	*Saimiri sciureus*	*Saguinus imperator*	*Saguinus fuscicollis*
Vertebrates	**6**	**6**	**4**	**2**	**13**
Frogs	1	1	1	2	3
Lizards	2	4	3		10
Birds	2		1		
Mammals		1			
Eggs	1	1			
Orthoptera	**5**	**14**	**34**	**57**	**61**
Lepidoptera	**17**	**12**	**50**	**26**	**6**
Larvae	4	6	30	16	3
Pupae	11	7	18	10	3
Adults	2		2		
Hymenoptera	**42**	**48**	**4**	**4**	
Ants	39	37	3	4	
Wasps	3	11	1		
Isoptera	**8**	**5**	**1**		
Coleoptera	**8**	**8**	**2**	**9**	**6**
Larvae	6	7		2	3
Adults	2	2	2	7	3
Miscellanea	**14**	**7**	**5**	**2**	**13**
Galls	1	1	2		3
Hemiptera	1	1	1		7
Millipedes		1			3
Snails	12	5	2	2	
Total no. of identified prey	178	178	175	49	31

NOTE: Values for each major prey category in boldface.

the same spectrum of prey types. *Cebus albifrons* is more partial to wasp nests and *apella* to snails, but those are the only differences worth mentioning, and they are minor ones. In both species, ants (actually, the brood of colonies) constitute the modal class, a fact that is directly attributable to their propensity to bite open hollow twigs and vine stems. The two leaf foragers, *Saimiri* and *Saguinus imperator*, likewise show similar prey distributions, with one important difference: *Saimiri* takes more lepidoptera (nearly all larvae and pupae), while *S. imperator* takes more orthoptera. The functional distinction here is that orthopterans are capable of escaping while lepidopteran larvae and pupae are not. With its stealthy advance and lightning attacks, *imperator* possesses a more effective technique for capturing mobile prey. *S. fuscicollis* also takes a large proportion of orthoptera, but these are almost invariably extracted from cavities and crevices rather than snatched from leaves. Curiously, of all the species, *fuscicollis*, the smallest, captures the greatest proportion of vertebrate prey, mostly lizards. The majority of these are seized on trunks as the animals scamper up and down, checking out every crack and fissure.

Intraspecific Comparisons

Since many of the interspecific differences just reviewed appear to be size related, it is reasonable to ask whether different-sized individuals of a given species manifest differences in their foraging behavior that parallel the interspecific trends. Pursuing the analysis to such depth requires both the ability to recognize individual group members in the field and large sample sizes. These requirements are best met in our data for *Cebus apella*, to which we shall therefore direct the inquiry.

The various recognizable age/sex classes of *Cebus apella* display an appreciable heterogeneity in their use of foraging substrates (Table 6.6; chi-square = 355, $p << 0.001$). Males search leaves and branches more than females and palms less; the subadult male opened leaves with exceptional frequency, though this may have been an individual idiosyncrasy; the infant avoided palms which require strenuous pulling and tearing, and concentrated more on leaves which are easily searched. It is noteworthy that sex differences seem to override age differences, as indicated by the greater resemblance of the patterns shown by juvenile and subadult males to the pattern of adult males than of adult females. This is a key result, for it denies that size is more important than species identity as a determinant of foraging behavior. Indeed, when we compare the patterns of juvenile *apella* and *Saimiri*, animals of the same size, we observe that the young *apella* emulate their elders very closely and do not deviate in any discernible way toward the behavior of the other species.

Table 6.6
Substrates searched by age and sex class in *Cebus apella* (percentage).

		Adult		*Subadult*		*Juvenile*	*Infant*
		Male	*Female*	*Male*	*Female*	*Male*	*Male*
Substrates searched	*No in. class*[a]	*3*	*3*	*1*	*1*	*2*	*1*
Leaf surface		26	19	26	25	17	30
Leaf interior[b]		19	17	31	17	24	18
Branch interior[c]		36	27	23	27	37	40
Palm crown		8	27	9	19	11	2
Other[d]		11	10	11	12	10	10
Sample size (N)		1186	1319	673	336	686	279

NOTE: Both adult and subadult females forage significantly more in palms than their male counterparts, $p << 0.01$.
[a] Average number in the study group over all samples.
[b] Refers to curled up live or dead leaves that were opened.
[c] Refers to twigs, vinestems, etc. that were stripped of bark or bitten open to expose cavity within.
[d] Includes foraging in vines, bamboo, cane, epiphytes, etc.

There are at least three possible interpretations of this result. One is that body size really is not a major determinant of foraging behavior, contrary to conclusions reached elsewhere in this chapter. Another is that young *Cebus* may simply be programmed to behave like adult *Cebus*. This would be difficult to disprove without controlled experiments. The third possibility is that young *Cebus* actually do emulate their elders, acquiring skills for a time when they can no longer run to their mothers for milk. At present we cannot distinguish between these alternatives, though there is some further evidence that bears on the matter.

While all *C. apella* are avid branch breakers, there is a size-related trend in this behavior in that smaller individuals bite open smaller branches (Table 6.7). As the principal rewards of this activity are the eggs and brood of ant colonies, it is reasonable to suppose that large-diameter branches in general harbor larger colonies. This being so, one would expect that all individuals would select larger branches if they were physically capable of opening them. But clearly this is not the case. Only 15% of all the branches opened by infants and juveniles were greater than 2 cm in diameter as compared to 44% of those opened by adult females and 53% of those opened by adult males. Thus there are size-dependent shifts in foraging behavior during the ontogeny

Table 6.7
Diameters of branches broken by different age and sex classes of *Cebus apella* (percentage).

	Branch diameter class (cm)						*Sample size (N)*[a]
	0–<0.5	*0.5–<1*	*1–<2*	*2–<5*	*5–<10*	*≥10*	
Infant	**57**	14	14	14	0	0	14
Juveniles	20	25	**41**	12	2	1	322
Subadults	14	23	**31**	20	7	5	133
Adult females	8	11	**37**	32	9	3	212
Adult males	3	13	30	**33**	14	6	262

NOTE: Modal values in boldface.
[a] Sample sizes generally lower than in other tables because it was not possible to estimate diameters of all branches broken by *Cebus.*

Table 6.8
Individual differences in substrate choice among 3 adult female *C. apella* (percentage of substrates searched).

	Substrate					
Female	*Leaf surface*	*Leaf interior*	*Branch interior*	*Palm crown*	*Other*	*Sample size (N)*
F_1	17	14	**29**	28	12	510
F_2	23	**24**	23	22	8	266
F_3	21	21	11	**35**	12	211

NOTE: Chi-square = 43.6, p<< 0.01; modal values in boldface.

of *C. apella*, but they are more quantitative than qualitative. Young *Cebus* do not pass through a *Saimiri*-like stage; instead they diverge both from *Saimiri* and from adults of their own species while nevertheless continuing to exhibit the same pattern of activities throughout their development. Whether the pattern itself is innate or learned by imitation remains a moot point.

Finally, we come to the question of individual variation within age and sex classes. Without belaboring the issue, I merely illustrate that such variability does exist in *C. apella* (Table 6.8). The three adult females in the study troop showed a highly significant heterogeneity of behavior. F_1 broke more branches, F_2 did more leaf foraging, and F_3 was most given to rummaging in palm crowns. What is not clear is whether these differences reflect individual preferences or the dominance structure of the troop. F_3, the most dominant of the females, was most partial to palms, the substrate favored by adult females as a group (Table 6.6). F_1 and F_2 then followed in their use of palms in accordance with their respective rank positions. These results are reasonable if one assumes that the most dominant animal chooses the preferred substrate, i.e., the one yielding the highest average rate of return (Morse 1970). While this may hold among females, it fails to account for the comparative disinterest in palms among adult males which are dominant to females (cf. Table 6.6).

Search and Capture Rates

The gross yield realized by a foraging animal is equal to the capture rate times the average caloric or nutritional value per prey. Unfortunately, because of the difficulty of estimating the size of many prey items, it was not possible to measure yields directly. Nevertheless, some noteworthy patterns can be discerned in the data on capture rates (Table 6.9). The number of searches per hour is not directly related to body size. The lowest rate is realized by one of the tamarins (*S. imperator*), the highest rate by *Saimiri*, and intermediate rates by the largest species (*Cebus* spp.). The differences are best explained by their divergent foraging tactics. *Saimiri* are compulsively active leaf foragers, searching superficially or with light manipulations. *Cebus*, as pointed out above, do a lot of heavy breaking and ripping open of substrates, operations that require more time than leaf searches and that consequently reduce the overall foraging rate. *Saguinus imperator*, with its scan and pounce tactic, spends more time looking at distant foliage and less time manipulating substrates; as a result it conducts a third as many searches per hour as *Saimiri*. Its rate of progression while foraging is also much less than that of *Saimiri*.

Saguinus fuscicollis goes to the opposite extreme from its congener in its expenditure of energy, since nearly all its searching is conducted on vertical surfaces. Instead of foraging continuously for long periods as do the other species, *fuscicollis* concentrates its searching in brief, intense bouts, usually on giant vine-draped trees. Upon arriving at such a tree, a group will begin by checking the axils of the buttresses; the animals then start upward, spiraling around the trunk or ascending thick liane stems. When they reach the crown, each individual normally goes out on a different major limb, checking every crevice and break in the bark. They are extremely agile in doing this, and often run along the undersides of large horizontal or inclined branches. After the tree has been thoroughly inspected, the group may reconvene on the main trunk and descend together, or the animals may jump at 40 m to adjacent large trees and work these from the top downward. Much of this, especially the branch searching in high crowns, is conducted at high speed, with indi-

Table 6.9

Search, success, and capture rates for leaves and other substrates (yearly means).

Species	No. searches[a]			Searches/hour[b,c]			Success rate[b,d]			Captures/hour[b,e]		
	Leaves	Other	Total	Leaves	Other	Total	Leaves	Other	Total	Leaves	Other	Total
Cebus apella	2093	2477	4570	31.2	36.3	67.5	0.43	0.41	0.42	14.2	14.6	28.8
Cebus albifrons	1994	1261	3255	40.7	26.2	66.9	0.37	0.38	0.38	15.6	10.1	25.7
Saimiri sciureus	3918	389	4307	81.8	10.0	91.8	0.60	0.76	0.61	48.5	7.5	56.1
Saguinus imperator	451	91	542	26.5	4.6	31.1	0.43	0.63	0.46	11.3	2.9	14.2
Saguinus fuscicollis	8	292	300	0.7	43.9	44.6	0.29	0.14	0.15	0.3	6.4	6.7

[a] All samples combined. Data from timed sequences only.
[b] Means of values from individual samples.
[c] Number of searches of the type indicated per hour of timed observations.
[d] Proportion of searches that resulted in captures.
[e] Product of the hourly search rate times the success rate.

viduals scampering quickly from knothole to knothole and making dramatic leaps from one major limb to another. Although *fuscicollis* spends less time foraging than any of the other species, the foraging it does undertake obviously requires an exceptional amount of exertion. Regardless of the frenetic activity, its prey capture rate (6 per hour) is by far the lowest of all.

Prey capture rates depend on the search rate and on the proportion of searches that result in successful captures. *Saimiri* leads in both these categories with the result that it captures prey at twice the rate of any other species, nearly one per minute. The *Cebus* species do distinctly less well in the percent of searches that lead to captures, and this is just as true when only leaf searches are considered. The reason for this is not clear. It may be that because of their larger size the *Cebus* rout more potential prey before reaching them. Or it may be that the *Cebus* are obliged to engage in redundant searches following the *Saimiri*, which typically precede them through the forest (see Chapter 8). The capture rate of *S. imperator* is about a quarter of that of *Saimiri*, owing to its reduced search rate and the fact that it makes more unsuccessful attempts on mobile prey.

Temporal Variation in Foraging Behavior

Diurnal patterns. Rather little is known about the diurnal activity patterns of arthropods in the tropical forest. It is easy to imagine, however, that changes in the ambient light or temperature regime could have pronounced effects on the microspatial distribution of small prey, or on the susceptibility to capture such prey. If so, one would expect predators to modify their hunting behavior in appropriate ways.

We have already examined the activity budgets of the five species and seen that foraging is conducted throughout the diurnal period. There is some reduction in the incidence of foraging in the early hours of the day (6:00 to 8:00); but this seems better explained by a tendency to feed on fruit at that time rather than by a disinclination to forage. This conclusion is borne out by data on success rates which show no pronounced diurnal pattern (Fig. 6.2). It appears that foraging effort is rewarded equally at any time of day. Nevertheless, there is a distinct tendency for the three larger species to adjust their foraging positions as the day progresses. Foraging is conducted relatively high in the morning and evening and lower during the middle hours of the day (Fig. 6.3). A similar diurnal pattern has been noted for insectivorous bird flocks in the Amazonian forest (Pearson 1971).

The pattern is indistinct or lacking in the tamarins. Compared to the other species, tamarins rise late and retire early, and do relatively little foraging at the beginning and end of the day. *S. imperator* confines its searching to a narrow zone between 3 and 10 m above the ground. Although the groups

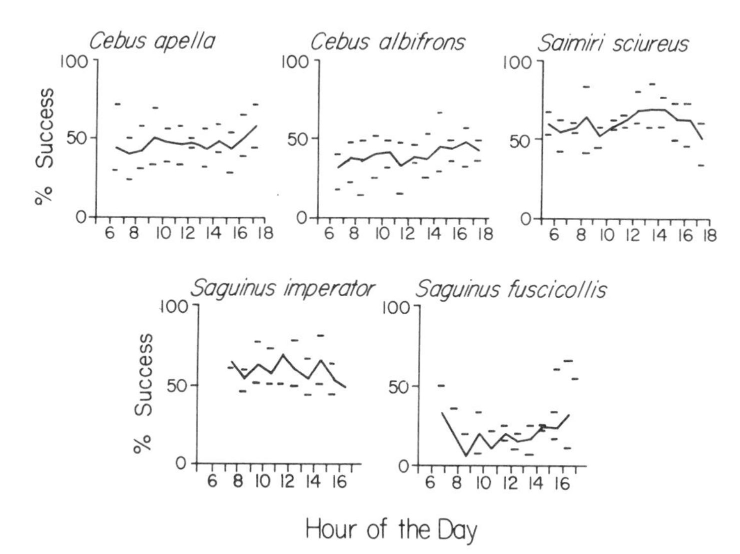

Figure 6.2 Success rate (percent of searches that resulted in capture of a prey item) as a function of time of day. Aberrant values in the first and last hours of activity are probably attributable to small sample sizes. Data from all seasonal samples combined.

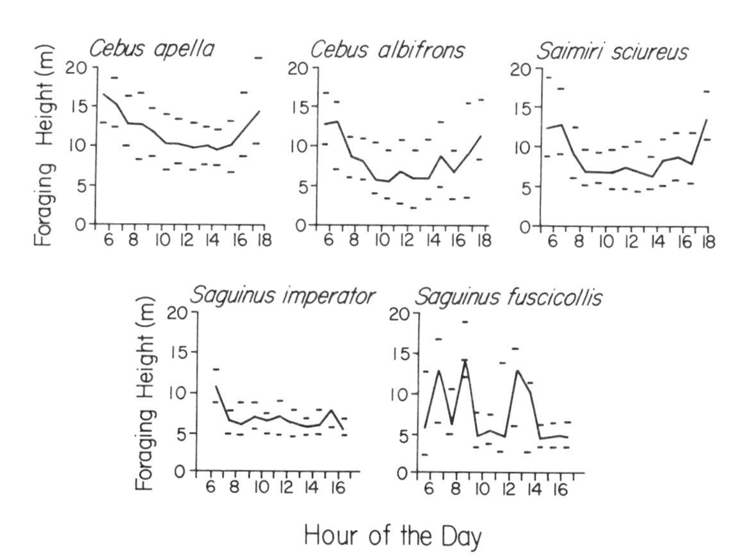

Figure 6.3 Height of foraging vs. time of day. Points are the means over all seasonal samples of the median values for each hourly interval. Bars indicate quartile ranges about the means.

ascend into the canopy to harvest fruit, they seldom remain there to forage. Possibly *imperator* stays low to keep a cover of foliage overhead in order to minimize exposure to aerial predators. *S. fuscicollis*, as previously described, searches trunks and branches at all levels between 1 and 45 m. The large vertical variance of this species thus obscures any possible diurnal pattern.

Two types of explanation come to mind for the vertical shifts in foraging shown by the *Cebus* species, *Saimiri*, and Pearson's bird flocks. One is that the predators are responding to some behavioral pattern on the part of their prey. Perhaps insects migrate downward at midday to avoid high temperatures and desiccating conditions in the canopy. Alternatively, it may be that insects are more alert and active in brightly illuminated crowns and thus more difficult to capture, thereby motivating their predators to move downwards. The latter possibility seems rather unlikely in view of our inability to detect any difference in success rates of high vs. low-foraging individuals at any time of day. Moreover, *Cebus apella*, which captures very few mobile prey, shows the pattern as distinctly as *Saimiri*, which captures a relatively high percentage of mobile prey.

The latter fact suggests that the behavior of the prey may not be the primary factor motivating the predators; this brings us to the second category of explanation. It may be that the predators themselves are reacting to external conditions that affect their success, or perhaps simply their comfort. Searching for prey is primarily a visual activity, and so it is probable that there is some optimal light intensity range in which prey are most readily detected. The light intensity in the understory in the early morning and late evening is very low, less than 0.1% of full sunlight, and may not be sufficient for keen discrimination. Hunting in the canopy at high noon might present other difficulties. There the contrasts between directly illuminated surfaces and shaded surfaces are extreme. Moving in rapid succession between bright and dimly lit sections of tree crowns would require constant adjustments of pupil diameter and the accommodation mechanisms, almost certainly to the detriment of hunting success. By moving to lower levels of the forest the animals can enjoy more uniform conditions (Papageorgis 1975), and can avoid the heavy heat and desiccation loads of the canopy as well. For the present, the evidence seems to point to this explanation of the pattern.

Seasonal patterns. The abundance, diversity, and taxonomic composition of arthropod communities are known to undergo strong seasonal variation in southern Central America at about the latitude of southeastern Peru (Robinson and Robinson 1970; Janzen 1973b; Smythe 1974; Ricklefs 1975; Buskirk and Buskirk 1976; Wolda 1978; Wolda and Fisk 1981). Of the localities that have been studied, the one that most resembles Cocha Cashu in seasonality and rainfall is Barro Colorado Island, Panama. A long-term environmental monitoring program there has included weekly night lighting for insects. The results of these measurements indicate a sharp increase in overall insect abundance coincident with the first heavy rains of the dry-wet transition period (May–June). The average catch then declines irregularly, reaching an annual low in the late months of the dry season (February–April) (Smythe 1974; Windsor 1976; Wolda 1978; Wolda and Fisk 1981). Within the broad seasonal march of changes in aggregate insect density there is of course a great wealth of underlying detail, as individual taxonomic groups wax and wane in importance in accordance with a variety of schedules. Such detail is beyond the scope of our analysis.

Ideally, we would have our own data on arthropod densities as a background against which to assess seasonal changes in the hunting behavior of the monkeys. Initially we did undertake a night lighting program; but due to the accidental loss of our battery supply, we were forced to abandon it half way through the year. In any case, the results were not especially gratifying in that the taxonomic overlap between the samples we did examine and the prey the monkeys were capturing concurrently was close to nil. Orthoptera were poorly represented, for example, and lepidoptera and ants were captured as adults or reproductives while the monkeys ate only immature stages. With such poor correspondence between what one would like to measure (prey actually available to monkeys) and what one was able to measure, the effort of carrying out the sampling was only marginally justified. Clearly what is needed is a way of sampling that mimics the kinds of searching operations the animals themselves perform. In lieu of such measurements, our only recourse is to draw inferences from the behavior of the animals.

Pronounced shifts in substrate use frequencies, for example, might be expected as a response to fluctuating success rates on alternative substrates. While this seems a reasonable type of possibility to anticipate, the indications of it in our data are rather weak and confined to the *Cebus* species (Table 6.10). The other three species are characteristically less versatile in their behavior and appear to follow similar habits throughout the year. But in both *Cebus* species there is a more or less marked shift away from leaf searching in the late rainy and early dry seasons. The trend is more extreme in *C. apella* (from 59% to 24% leaf searches), the reduced interest in leaves being compensated by greater efforts devoted to breaking branches and rummaging in palm crowns. In *C. albifrons* the shift is rather slight (ca. 65% to ca. 55% leaf searches) and accompanied by minor increases in palm foraging and other activities (bamboo, cane, etc.).

Looking now at *Cebus apella* in more detail, we might expect to see a reduced success rate on leaves and/or an increased success rate on branches and palms in the dry season (June), when leaves are used sparingly (Table 6.11). But the situation is really not so simple. Recalling from the Panamanian results described above that arthropod abundance can be expected to peak

Table 6.10
Seasonal variation in substrate use (percentage of searches).

		Substrate				
Species	*Sample*	*Leaves*	*Branches*	*Palms*	*Other*	*Sample size (N)*
Cebus apella	Aug.	51	36	6	7	373
	Oct.	59	31	4	6	1909
	Jan.	59	27	11	3	1291
	April	35	42	20	3	222
	June	24	54	17	5	1613
Cebus albifrons	Aug.	63	30	2	5	259
	Dec.	66	25	2	7	1857
	March	57	26	4	13	453
	June	53	29	7	11	1066
Saimiri sciureus	Sept.	93	5	0	2	1817
	Feb.	81	12	0	7	313
	May	90	6	$<<1$	4	1398
	July	89	6	0	5	1109
Saguinus imperator	Aug.	95	4	0	1	104
	Oct.	88	10	0	2	49
	April	81	13	0	6	373
	July	87	10	0	3	156
Saguinus fuscicollis	Nov.	5	28	0	66	113
	April	3	22	0	75	288
	July	<1	31	0	69	108

near the beginning of the rainy season, we can take the October sample as a frame of reference. Comparing June to October we find small but consistent decreases in the success rate in all 7 substrate categories. There is little change in the relative success on leaves vs. branches vs. palms, but overall there is a substantial decrease in the number of captures per hour of foraging. These data are contained in Table 6.12. Both *Cebus apella* and *C. albifrons* show high capture rates in the early rainy season, and much lower rates at other times. While it seems likely that the shift in capture rates is most directly a reflection of altered prey abundance, there are some other possibilities that merit consideration.

One is that changes in the catch rate result from shifts in emphasis from one substrate type to another. Some substrates can be searched quickly (leaves) while others (e.g., palm crowns) require extended effort. If the average reward per capture were greater in palms at certain seasons, a switch to palm foraging

Table 6.11
Seasonal variation in substrate use, substrate-specific foraging rates, and success rates in *Cebus apella*.

	Break branch			*Break vine stem*			*Palm crown*			*Live leaves*			*Dead leaves*			*Live palm leaves*			*Dead palm leaves*		
Sample	*N*	*N/hr.*[a]	*Suc.*[b]	*N*	*N/hr.*	*Suc.*	*N*	*N/hr.*	*Suc.*	*N*	*N/hr.*	*Suc.*	*N*	*N/hr.*	*Suc.*	*N*	*N/hr.*	*Suc.*	*N*	*N/hr.*	*Suc.*
Oct.	265	12.4	0.40	76	3.6	0.57	103	4.8	0.41	570	26.7	0.66	164	7.7	0.43	232	10.9	0.52	57	2.7	0.48
Jan.	169	11.3	0.36	75	5.0	0.33	141	9.4	0.32	338	22.6	0.40	82	5.5	0.33	155	10.4	0.27	50	3.3	0.38
Apr.	44	12.6	0.36	10	2.9	0.20	56	16.0	0.31	28	8.0	0.54	10	2.9	0.11	27	7.7	0.23	7	2.0	0.57
June	550	21.5	0.35	220	8.6	0.33	310	12.1	0.39	117	4.6	0.63	94	3.7	0.30	97	3.8	0.28	65	2.5	0.37

[a] Number of searches of the type indicated per hour of timed observations.
[b] Proportion of searches that resulted in captures.

Table 6.12

Hourly search, success, and capture rates for all species and samples.

Sample Species	Early wet (Oct.–Dec.)				Late wet (Jan.–Apr.)				Early dry (May–June)				Late dry (Jul.–Sept.)			
	Hrs.[a]	Sear./hr.[b]	Suc.[c]	Cap./hr.[d]	Hrs.	Sear./hr.	Suc.	Cap./hr.	Hrs.	Sear./hr.	Suc.	Cap./hr.	Hrs.	Sear./hr.	Suc.	Cap./hr.
Cebus apella	21.34	79.1	0.57	44.9	14.97	72.9	0.37	26.7(Jan.)	25.62	62.3	0.38	23.4				
					3.49	55.6	0.36	20.1(Apr.)								
Cebus albifrons	18.16	96.0	0.41	39.7	10.69	40.0	0.35	13.9	16.74	64.7	0.37	23.6				
Saimiri sciureus	17.45	89.4	0.62	55.3	3.79	79.4	0.67	53.5	14.91	91.3	0.56	51.4	10.14	106.9	0.60	64.1
Saguinus imperator	1.39	30.9	0.38	11.7	13.44	25.2	0.50	12.7					4.31	37.1	0.49	18.3
Saguinus fuscicollis					5.66	41.5	0.13	5.4					1.36	47.8	0.17	7.9

[a] Number of hours of timed foraging sequences in sample.
[b] Number of searches of all types per hour of timed observations.
[c] Proportion of searches that resulted in captures.
[d] Product of the hourly search rate times the success rate.

might be accompanied by decreased search and capture rates with no loss in the nutritional yield per unit time. Such an effect could be part of the story, but it is doubtfully more than a small part. There is a sharp (40%) drop in the capture rate of *C. apella* between the October and January samples without any corresponding decrease in the frequency of leaf searches and only a modest increase in palm use (Table 6.10). The major shift in substrate use came between the January and April samples, but the associated decrease in catch rate (25%) was less than the previous drop. The same can be said even more emphatically of *albifrons*, which experienced a 3-fold decrease in catch rate between the December and March samples, accompanied by only minor adjustments in substrate use. We thus conclude that most of the variation in capture rates shown by the two *Cebus* species is independent of seasonal patterns in the choice of substrates. The same conclusion applies with even greater certainty to the other species.

If seasonal variations in the capture rate cannot be attributed to shifts in substrate use, how then can they be explained? Capture rates are computed as the product of the search rate (number of searches per hour) times the weighted mean success rate over all types of searches. All observed variation in capture rates can thus be explained by variation in the underlying search and success rates.

Let us consider the effect a change in prey abundance might have on search and success rates. If an animal were foraging blindly, that is, searching leaves or other substrates purely at random, a change in prey abundance would translate directly into a change in success rate. If, on the contrary, an animal were visually scanning its environment for signs of prey (curled leaves, damaged foliage, or other telltale indications), and responding only to such signs in initiating searches, a change in prey abundance would translate directly into a change in the frequency of telltale signs and hence into a change in search rate. We have thus arrived at a dichotomy of predictions that depends on whether or not a predator initiates searches in response to signs of prey.

Now we can apply this reasoning to the five primates. Two of the species seem to do a great deal of blind foraging—*Cebus apella* in its breaking of dead twigs and rummaging in palm crowns, and *Saguinus fuscicollis* in its tactile inspection of the dark interstices of tree cavities. The other three species are predominately leaf foragers and could be expected to respond to insect signs in launching searches. We might therefore predict that the first two species would show relatively more seasonal variation in success rates while the latter three would show more variation in search rates. This is indeed what we find (Table 6.13). It thus appears that the leaf foragers are not manipulating substrates at random but are reacting to stimuli that suggest the presence of prey. One can infer that the probability of a prey item being associated with a given kind of stimulus remains relatively constant. It is

Table 6.13
Relative magnitudes of variance and covariance in search and success rates as they contribute to seasonal variation in capture rates.

	Seasonal variance	Proportion of variance contributed by			
Species	*in capture rate*	*Search rate*		*Success rate*	*Interaction*
Cebus apella	92.1	.205	<	0.389	+0.407
Cebus albifrons	113.2	.689	>	0.039	+0.276
Saimiri sciureus	23.4	1.646	>	0.599	−1.217
Saguinus imperator	8.4	.627	>	0.455	−0.092
Saguinus fuscicollis	1.6	.134	<	0.407	+0.468

mainly the frequency of stimuli that varies seasonally. But in the case of dead twigs or knotholes, it appears that the stimulus is the substrate itself, and that variation in capture rates is due to variation in the frequency with which prey are associated with the substrate.

The interaction terms are not interpreted so simply. In four out of the five species they are small or positive. Positive interactions indicate that success rates were high when search rates were high and vice versa. This would suggest an increased association of prey with signs of prey during periods of high prey density. *Saguinus imperator* shows a very small interaction term, as would be expected in a species that often attacks prey directly. Curiously, *Saimiri* shows a strongly negative interaction, indicating that success rates were low when search rates were high. This could result from changes in the inherent catchability of prey at different seasons, such that prey were most difficult to capture when sign of them was most abundant. But if this were true for *Saimiri*, why would it not also be true of *C. albifrons*, which instead shows a moderately large positive interaction term? There is yet another possibility. A negative interaction between search and success would result if low success rates spurred the animals to greater efforts. This might actually occur if the animals' greatest incentive coincided with a period of low success rates. It may be that such a nexus exists in the early dry season when alternative foods are scarce. Inspecting the data more closely, we note that *Saimiri* search rates were highest in May and July, but particularly in July when we will recall, in the absence of any fruit in the study area the animals were doing little else than foraging (see Chapters 4 and 5). The success rate in this sample was slightly below average, but the capture rate was the highest recorded.

All of the excess over the mean capture rate is accounted for by an accelerated search rate. *Saimiri* may simply work harder when hungry.

The amount of seasonal variation in capture rates is not the same in all species, nor could one expect it to be unless all species were exploiting the same prey resources in the same ways. Nevertheless, there is a common phenological pattern in the results (Table 6.12). All the species show relatively low capture rates in the late wet and early dry seasons, and relatively high ones in the late dry and early wet seasons. This is about what we would have anticipated from the seasonal trends in the Panamanian night-lighting results described above. The one exception to an otherwise consistent pattern is the short October *S. imperator* sample which shows a low capture rate relative to April. Possibly this is an artifact of the small sample size.

The amplitudes of the seasonal variation in capture rates range from a maximum of 2.9-fold in *Cebus albifrons* to a minimum of 1.2-fold in *Saimiri*. Why these two should represent the extremes, when their foraging is similar in so many ways, is unclear. The average amplitude is around 2-fold. It would be interesting to know how closely this reflects the actual seasonal variation in arthropod abundance, but the answer will have to await the results of future research.

Discussion

Implications of Body Size for Insectivory

For the sake of argument, let us consider how the five species might fare as obligate predators. Although we shall not attempt to estimate the biomass of prey that a typical individual of each species would need for self-maintenance, we can easily infer their relative needs from their respective weights. Assuming that metabolic needs scale as the 0.75 power of body weight (McMahon 1973), it is easily calculated that a *Saimiri* would need to catch about twice the mass of prey per day as a *Saguinus fuscicollis*, and that a *Cebus* would have to catch about 4.5 times as much. Given the capture rates we have just examined, these expectations clearly are not being met unless the larger species are catching a highly disproportionate number of large prey. However, the available data, such as they are, indicate the contrary (Table 6.14). *Cebus* spp. catch only half as many prey per day as *Saimiri*, and only about a quarter as many large prey. The disparity is undoubtedly less than the data imply because a certain number of the ant broods consumed by *Cebus* should qualify as "large" prey. Nevertheless, it is hard to imagine that the biomass consumed by an average *Cebus* per day is any more than that consumed by a *Saimiri*, much less 2½ times as much as would be required by its larger body size.

Table 6.14
Estimated per day catches of all prey and large prey.

		All prey		*Large Prey Only (≥ 1 cm)*		
Species	*Mean no. hours per day in foraging*	*Mean capture rate (no./hr.)*	*Est'd no. prey per day*	*Mean capture rate (no./hr.)*	*Percentage recorded as large*	*Est'd no. large prey per day*
Cebus apella	5.9	28.8	170	0.34	1	2.0
Cebus albifrons	4.7	25.7	121	0.86	4	4.0
Saimiri sciureus	5.9	56.1	331	2.5	5	14.7
Saguinus imperator	3.3	14.2	47	1.9	15	6.3
Saguinus fuscicollis	1.5	6.7	10	2.5	42	3.8

It follows that *Cebus* could not survive as an insectivore no matter how diligently it were to hunt.

Although the tamarins spend considerably less time foraging than any of the larger species, their daily yield of large prey is as good or better than that of *Cebus*. This results mainly from the fact that they capture a much higher proportion of large prey, a tendency that is especially pronounced in *Saguinus fuscicollis* (Yoneda 1981).

Normally, predators prefer large prey so long as the handling time does not become excessive (Krebs 1978). Consequently, average prey size is generally found to increase with the size of the predator, regardless of whether the predators are fish (Werner 1977), lizards (Schoener 1967, 1968), birds (Storer 1966; Ashmole 1968; Hespenheide 1971), or mammals (Schaller 1972). Instead of this common and intuitively simple pattern, the results in Table 6.14 suggest that average prey size for the five monkeys decreases as the body size of the predator increases. We have here a paradoxical result that contradicts one's sense of propriety about matters of optimal foraging. Surely *Cebus* do not prefer small prey, for on rare occasions they capture fully grown rats, opossums, lizards, or fledgling birds and devour them with evident relish. The fact that they do not routinely capture large prey can be interpreted reasonably only to mean that they are unable to do so. We infer that they are handicapped by their large body size. The unavoidable commotion produced by their movements along and between branches renders it virtually impossible for them to get within striking distance of mobile prey (which include most large prey, such as the adults of insects and vertebrates). Having no other alternative, they put their size to best use by specializing in destructive foraging. Even though their efforts are not well rewarded in relation to their gross metabolic requirements, the yield is presumably good enough to satisfy their needs for protein, and perhaps for certain minerals as well.

The smaller monkeys face a different set of options. Destructive foraging, given its relatively low payoff, is apparently not worth the effort in the face of an improved likelihood of catching large prey, for *Saimiri* and *Saguinus* do almost none of it. Indeed, as the chances for obtaining large prey improve, a predator should modify its hunting practices to maximize its opportunity to catch them. The advantage of doing this is apparent when one considers that a single 5-cm orthopteran provides the nutritional equivalent of scores of prey in the 0.5–1.0 cm range. With a very low probability of obtaining large prey, *Cebus* unselectively take what they find. In contrast, the tamarins appear to be highly selective, or at least one would so infer from the greatly increased proportions of large prey in their catches.

Selectivity should be exercised to the extent that harvesting small prey jeopardizes the probability of catching large prey. This could occur through increased agitation of the substrate and/or through decreased attentiveness on the part of the predator. One can readily understand how selectivity is favored in *S. imperator* with its stalk and pounce hunting technique. In the case of *S. fuscicollis*, the high frequency with which it captures large prey seems to result from a prevalence of such prey in holes and crevices that are big enough to accommodate a monkey's hand.

We have arrived at a rationale for the prey-size selectivity of *Saguinus* and the lack of it in *Cebus*, but what about *Saimiri*, which is intermediate in both size and behavior? *Saimiri* take considerable numbers of large prey by virtue of their high overall capture rate, but the proportion of large prey is barely greater than in *Cebus albifrons*. *Saimiri* do not stalk, and as a consequence probably flush many items that *S. imperator* would capture. Another major factor is that *Saimiri* forage in large mixed troops. Trailing individuals are frequently searching along arboreal pathways that had just been traversed by another individual. The great general commotion, which is audible nearly 100 m away, flushes numerous large insects and small vertebrates that are chased down by several species of commensual birds (see Chapter 8). It thus seems fairly certain that many of the most desirable prey items escape or are snatched up by birds before the monkeys ever get within striking range. Being too small to engage effectively in destructive foraging, and traveling in groups too large to permit the stalk and pounce technique, *Saimiri* attempt to maximize their capture rate by speeding up the rate of searching and taking prey unselectively as encountered. By speeding up (three times as many searches per hour as *S. imperator*), they no doubt flush some insects *imperator* might catch, but they also harvest four times as many prey per unit time.

From the data in Table 6.14 one might guess that *Saimiri* and the *Saguinus* species are meeting roughly similar proportions of their metabolic needs with prey, though *Saimiri* take a much greater part of each day to do it. All three species catch large prey at about the same rate, but being smaller, the *Saguinus* do not need so many per day. So, while *Saimiri* continue their tireless hunt, *Saguinus* seek safe shelter for rest and socializing. In the jargon of optimal foraging theory, *Saimiri* can be considered an energy maximizer, while *Saguinus* spp. are time minimizers (Schoener 1971). *Cebus* spp. spend a great deal of time in foraging in the manner of energy maximizers; but in relation to their size, the yield seems to fall far behind that of the smaller species.

The role of body size in insectivorous primates is thus extremely critical. It determines which of several alternative hunting modes is most profitable:— stalk and pounce, cursory searching, or destructive foraging. By exercising strong constraints on foraging methods, body size indirectly determines the effectively available prey supply, including both its size distribution and taxonomic composition. But most crucially, the observed failure of prey capture rates to scale with the mass of the predator implies that there must be an abrupt upper limit on the practicality of insectivory as even a supplementary mode of life. This limit is encountered when an animal's protein requirement can no longer be satisfied by the rate at which it captures prey. Our results suggest that this limit lies not much above the size of an adult *Cebus*.

Interference vs. Exploitation Competition in Resource Partitioning

Competition for fruit and competition for small prey lead to radically different methods of resource partitioning. Fruit is generally available to all comers and, once located, harvesting it requires no great specialization of morphology or technique. Among frugivorous birds there may be a strong positive relationship between the size of the bird and the average size of the fruits eaten, but birds are in general much smaller than monkeys and are required to swallow their food whole (Diamond 1973). Among our five primates there is no correlation between body size and the size of fruits eaten, as was demonstrated in the preceding chapter. Thus it is clear that the significance of body size for harvesting fruit vs. insects must be very different.

Differential exploitation of or access to fruit crops can be achieved through preemption or interference. Preemption implies doing something first and in this context refers to the ability to consume fruits at an earlier stage of ripeness than another species. We frequently observed, for example, that certain trees, usually figs, were used by *Callicebus* or *Alouatta* a number of days before they began to attract large numbers of *Cebus*, *Saimiri*, and *Ateles*. Frugivores can thus specialize in keeping one step ahead of their competitors, a practice that probably requires adaptive modifications of the digestive process but not of any obvious features of external morphology.

Interference competition, in contrast, is affected by aggressive intimidation, the larger or more belligerent species successfully driving off a less assertive user from a contested resource. This kind of interaction is very prevalent in primates (Waser and Case 1981), and we observed numerous instances of it both between and within species. A band of *Ateles* can rout a troop of *Alouatta*, for example, and *Cebus albifrons* can dislodge *Cebus apella* from a *Scheelea* palm; either *Cebus* expels *Saimiri* at will from small fruit trees (but not large ones), and *Saguinus imperator* routinely intimidates and supplants *S. fuscicollis*. Priority of access to especially desired resources is predetermined by an interspecific dominance hierarchy based largely but not entirely on size. Low-ranking species can obtain a share of contested resources by preemption or by waiting their turn, but the latter alternative is precluded if the dominant species consumes all the available ripe fruits. A subordinate species can also avoid intimidation by seeking resources that are not preferred or that are too thinly scattered to be efficiently harvested by a larger dominant. It is this latter tactic that has been developed to a striking degree in *Saguinus*.

Competition for small prey is mediated through entirely different means. Unlike fruit, which is concentrated in discrete units, small prey are scattered and hidden. It is impossible for a dominant species to control access to prey because the resource is effectively everywhere. Neither is it likely that differences in preference for particular prey taxa will confer much ecological separation, for such selectivity is generally counterproductive for a searching predator. Here we have a case of pure exploitation competition in which the activities of each species effectively reduce the prey capture rate of all the others. Optimal foraging theory tells us that in this situation there should be specialization on patch type, not on prey taxa (MacArthur and Pianka 1966; Charnov 1976a).

It is not entirely clear here what is meant by a patch or discrete type of "habitat." One definition, and this is the one the authors most likely had in mind, is that a habitat is a distinct type of vegetation. Following this view we might expect to see a pattern of habitat selectivity in the choice of foraging areas by the five primates. But no such pattern is apparent. However, there is another plausible interpretation of "habitat" or "patch" as used by MacArthur and Pianka, and that is for it to refer to distinct substrates. The extraordinary structural complexity of the tropical forest permits the evolution of many kinds of specialized foraging behaviors, as has been extensively documented for birds (Orians 1969b; Wiley 1971; Pearson 1977; Terborgh 1977, 1980). A species that foraged high might just as well be regarded as being in a different ''patch'' from one that foraged low—as one that foraged high in an adjacent type of vegetation. By extension of this reasoning, a species that foraged on trunks could be considered to be using a different patch from one that foraged in leafy branches. The main point is that species which are engaged in exploitation competition should separate their activities spatially, while those that are engaged in interference competition have no recourse but to separate their activities temporally, assuming of course that the competing species do not diverge onto different resources altogether. Our observations are basically in accord with these expectations. Many of the documented interspecific differences in foraging behavior can be viewed as types of spatial separation, i.e., differences in substrate use, perch type, and vertical foraging position. Differences in capture technique and taxonomic composition of the catch, as previously stressed, are best regarded as secondary consequences of the spatial aspects of foraging. The one result that clearly was not anticipated by previous theoretical musings was the negative correlation between body size and capture of large prey. This is clearly a peculiarity of the arboreal environment in which a large animal cannot easily move without causing vibrations that can be detected by potential prey.

My final comment concerns the role of body size as it relates to the two types of competition: interference and exploitation. Large body size is obviously advantageous in conferring dominance in aggressive encounters; and although it carries a cost in the form of increased metabolic demand, under appropriate circumstances the cost could easily be offset by improved access to major resource concentrations. However, the very feature (large size) which enhances a species' success as a frugivore is highly detrimental to its success as an insectivore. It is thus clear that omnivores are faced with an adaptive trade-off between the antithetical goals of becoming better interference competitors or better exploitation competitors. It is this adaptive dichotomy that may lie at the evolutionary roots of the major size differences we find in the omnivorous New World primates.

Summary

This chapter presented comparative data on the prey catching (foraging) habits of the five species. We recorded the search and capture movements of actively foraging animals in timed intervals. Yearly sample sizes ranged from 7.02 hours and 297 substrates searched in *S. fuscicollis* to 65.42 hours and 4,570 substrates searched in *C. apella*.

Patterns of substrate use vary markedly between the species. *Saimiri* and

Saguinus imperator are leaf foragers, directing between 85 and 90% of their searches to this one class of substrate. In contrast, *Cebus apella* specializes in destructive foraging, such as breaking open hollow twigs and branches, rifling through the debris in palm crowns, and stripping bark from dead trunks and limbs. *Cebus albifrons* is intermediate in its behavior, engaging in a great deal of superficial leaf foraging as well as in less strenuous kinds of destructive foraging. By virtue of its skill in cling-and-leap locomotion, *Saguinus fuscicollis* proves to be an extreme specialist, doing nearly all its foraging on broad, vertical surfaces (trunks, major limbs) where it finds its principal substrate, knotholes (62% of all searches).

The perches selected by the animals during foraging also differ. The modal class for *C. albifrons, Saimiri*, and *S. fuscicollis* consisted of trees, while for *S. imperator* it was vines and for *C. apella*, palms. The frequency of use of vines as supports decreased with the size of the animal, with the exception of *S. fuscicollis* which, as already mentioned, forages mainly on bark. *C. albifrons* does more ground foraging (10%) than any of the other species.

Which of several capture techniques was used was found to depend on whether the prey was potentially mobile or not and on whether it was exposed or hidden. All of the species used at least four of the five recognized capture modes, but the proportions in which they were used varied widely. *Saimiri* and *S. imperator* mostly captured exposed prey (⩾70%), while the other species mainly captured hidden prey (⩾75%). Differences in search and capture behavior appear strongly to influence the taxonomic composition of the catch. The modal prey taxon for both *Cebus* species was hymenoptera, principally the brood of ant colonies obtained by breaking open hollow twigs. Half the catch of *Saimiri* consisted of lepidoptera, 96% of which were in the larval or pupal stages, and hence can be considered immobile prey. The tamarins, in contrast, concentrate on mobile prey, as about 60% of the identified items were orthoptera. Curiously, *S. fuscicollis*, the smallest species, captured the highest proportion of vertebrate prey (13%), the majority of which were lizards.

The possibility of intraspecific differences in foraging behavior was investigated with *C. apella*, as all the members of the study troop were individually recognizable. Minor quantitative differences are apparent in the use of various substrates by the several age and sex classes, the most notable contrast being that females of all age classes forage more in palms than do males. The search behavior of young *Cebus* unambiguously resembles that of adult *Cebus* rather than that of equal-sized *Saimiri*.

There is little vertical separation of the foraging activities of the five species, in contrast with reports on the ecological distinctions between sympatric Asian and African monkeys. Nevertheless, three of the species (*C. apella, C. albifrons*, and *Saimiri*) undergo a pronounced diurnal "migration" similar to that reported for insectivorous bird flocks in the Amazonian forest. The animals forage high in the early morning and late afternoon, and in the lower and middle stories of the forest throughout the rest of the day. I speculate that this is a response to ambient light levels in which the animals seek conditions that most facilitate discrimination of their prey against the background.

Prey capture rates differ between species and between seasons. *Saimiri* consistently take prey at the highest rate, about one per minute averaged over all seasons. *Cebus* have intermediate capture rates of about one every two minutes, while the tamarins make captures even less frequently. However, the low capture rates of the tamarins are compensated by their skill at obtaining large prey (≥1 cm), which they do with far greater consistency than any of the other species. This contrasts strikingly with the usual situation in which larger predators take larger prey. The paradox is explained by the fact that most large prey are mobile. The greater commotion produced by the larger monkeys as they move through the trees prevents them from getting within striking distance of the most desirable items such as lizards, frogs, orthoptera, cicadas, etc. The tamarins avoid disturbing their quarry through hunting by stealth (*S. imperator*) or by extracting them from their daytime hiding places (*S. fuscicollis*).

Seasonal variation in capture rates and in substrate use patterns are interpreted as responses to fluctuations in prey abundance and in prey dispersion in the vegetation. Capture rates can be decomposed as the product of search rates times success rates. We found that seasonal variation in the capture rates of the leaf foragers (*C. albifrons, Saimiri*, and *S. imperator*) correlated more strongly with variation in search rate than with success rate. This was interpreted to mean that searches are not conducted at random but in response to some perceived sign of insect activity or presence, e.g., rolled leaves, damaged foliage, etc. Quite the opposite was true of the species that do most of their foraging blind, *C. apella* (hollow twigs, debris in palm crowns) and *S. fuscicollis* (knotholes). In these, variation in capture rates was better explained by variation in success rates, as the proportion of substrates that contained prey varied in accordance with overall prey abundance.

The discussion considers the adaptive significance of body size as it influences the mechanisms of resource partitioning in insectivorous primates, and as it relates to the evolution of improved ability to compete via interference or exploitation of diffuse resources.

7 Ranging Patterns

It is possible to imagine several ways in which the harvesting of essential resources, or some other important life function, could lead to distinct patterns of spatial utilization. A species that depended on uniformly distributed, self-renewing resources, for example, could be expected to cover its territory or home range more or less evenly (Cody 1971; Pyke et al. 1977). At the other extreme, a species that relied on highly patchy resources would use space in a very irregular fashion, shifting the focus of its activities from one spot to another as a succession of patches went through periods of production and decline. Exigencies other than the search for food can and do impinge on movement patterns. Use of a traditional roost site or nest, for example, transforms the problem of optimal utilization of space. This variant has been investigated theoretically in what has become known as the problem of central place foraging (Horn 1968; Anderson 1978; Orians and Pearson 1979). Another variant, representing the contrary situation, could arise in a species that devoted a great deal of time and energy to defending its territorial boundaries. In this case the use would be more concentrated around the margins than in the center of the territory. These are four possible models of home range use, leading respectively to uniform coverage, to coverage that is focused on shifting patches, or to coverage concentrated at the center or around the periphery of the area. Each pattern reflects the influence of a dominant factor in the life of a species.

While these models will be helpful in guiding our thinking, and sometimes in suggesting interpretations of empirical results, the real world is considerably more complex than is indicated by any of them. One reason for this is that any given excursion may serve two or more functions. Insect foraging can and often is conducted during transits between fruiting trees, and territorial patrolling can at the same time serve an exploratory function in locating new resource patches. In such cases the function cannot be assigned unambiguously to either of the two alternatives. It is this frequent overlay of functions that makes the interpretation of movement patterns difficult.

Even if we assume that most travel is related to resource harvesting, and that other requirements are served along the way, it would be naive to expect monkeys to show any simple pattern of random or systematic coverage. It is obvious that their behavior is more complex. One factor is the tendency to return to resources that are not completely depleted by one visit, e.g., many fruit trees. Areas in which insect foraging is especially rewarding could be another example. Such a focus of attraction could introduce some semblances of central place foraging for as long as the resource lasted. But if several major resources are being exploited contemporaneously—a situation that is closer to the usual reality—then there are several "central places" in effect at once. Now, if new resources are being discovered, and old ones are being abandoned on a daily basis, the situation gains still further in reality, but becomes too cumbersome to model effectively.

At a more practical level, a major difficulty in the interpretation of animal movements is that the parameters we are able to measure in the field—rate of progress, distance traveled per day, area covered, etc.—are not directly related to what is accomplished by the travel. Such measures serve to quantify movement, but they tell us little about its function. Nevertheless, by knowing what resources are important to the animals, and something about how these resources are distributed in space, we may be able to judge the degree to which movements are related to the procurement of different components of the diet. Differences between species in the use of space can provide insights, as can seasonal shifts in the behavior of single species.

We shall proceed then by comparing the ranging behavior of the five species, after which we shall look more closely at each one to see whether the behavior changes in response to high vs. low availability of fruit. Finally, the ranging habits of our species will be contrasted with those of a number of other species for which data exist in the literature in an effort to reach some general conclusions about the relations between diet and the use of space.

Method

To document the daily movements of the study troops, we developed what we termed the follow-and-map technique. At the start of a move, the observer took a compass reading on the direction taken by the troop and followed it, counting paces. Usually a tree or other landmark in direct line with the compass course was used as a directional guide. Then, when the landmark was reached, or when the monkeys began to deviate from their initial trajectory, the observer stopped, recorded the distance moved and the time, and took another compass sighting. Each observer calibrated his or her own paces so that the lengths of moves could be recorded directly in meters. With practice we were able to apply this procedure with enough precision so that repeated estimates of a given location (e.g., a fruit tree approached on various occasions from different directions) usually mapped within 20–40 m of its actual position (as later confirmed by direct compass and tape measurement). We obtained additional reference points in each day's route every time the animals crossed a trail. The observer simply paced to the nearest location tag, recorded the position and continued with the monkeys. Later, the compass sightings, distances paced, and times were charted on 1:5,000 scale maps of the study area to provide accurate records of the daily meanderings of the troops, and of the positions of all fruit trees visited.

Home Range Characteristics of the Five Species

Quantitative Comparisons

Troop home ranges of the five species vary over a factor of nearly ten, from 30 ha to over 250 ha, notwithstanding their similar trophic status (Table 7.1).

Table 7.1
Home range and population density statistics for five omnivorous primates.

Species	*Mean troop size*	*Population density[a] (no./km²)*	*Biomass (kg/ha)*	*Home range size[b] (ha)*	*Home range per individual[c] (ha)*	*Area per individual[d] (ha)*	*Overlap index[e]*
Cebus apella	10	40	1.0	80	8.0	2.5	3.2
Cebus albifrons	15	35	0.8	⩾150	>10.0	2.9	⩾3.5
Saimiri sciureus	35	60	0.5	>250	>7.1	2.0	>3.6
Saguinus imperator	4	(12)[f]	0.05	30	7.5[g]	7.5[g]	1.0
Saguinus fuscicollis	5	(16)[f]	0.05	30	6.0[g]	6.0[g]	1.0

[a] From a strip census survey of the entire study area (Janson 1975).
[b] Minimal annual home ranges occupied by the study troops of each species.
[c] Area of home range divided by number of individuals per troop.
[d] Inverse of population density.
[e] Home range per individual divided by area per individual. The values roughly indicate the number of ranges that overlap any point in the study area.
[f] The estimates given in parentheses are those derived from a census taken in 1974 (footnote a). The area censused was fully occupied by *Saguinus* territories. However, much of the peripheral part of the study area appears not to be occupied by *Saguinus* (see Fig. 7.10).
[g] Values based on population densities in occupied habitat.

As already noted, however, there are great differences among the species in body size and in the average number of individuals per troop, factors that could help to mitigate the large discrepancies in spatial utilization. By dividing the average home range size by the mean number of individuals per troop, one obtains an estimate of the per capita area within the home range (home range area per individual). Surprisingly, these values now show a strong consistency across all the species, varying within the narrow range of 6 ha for *Saguinus fuscicollis* to 10 ha for *Saimiri*. That these values should be so similar is rather puzzling, since the sizes of the animals are so different.

There is a complication in the data that needs to be eliminated before we proceed. It results from the fact that there is a considerable amount of overlap in the home ranges of the larger three species, but not in the rigidly territorial *Saguinus* spp. This can be controlled by simply taking the inverse of population density, which gives estimates of the area per individual. We now emerge with the even more unlikely result that the two *Saguinus* species require twice or three times as much area on a per capita basis as do the larger species. This is particularly anomalous in view of their small body sizes, for it means that their biomass densities are on the order of 10 to 20 times less than for the other species. Although this poses an interesting problem in itself, it is not one we shall be able to resolve in the present work.

An important issue in the understanding of spatial utilization patterns is that of home range overlap—why it should be extensive in one species and not in another. A very simple index of overlap is obtained by dividing the home range per individual (Table 7.1, column 5) by the total area per individual (Table 7.1, column 6). For rigidly territorial species such as *Saguinus* which show no overlap, the index has a value of 1.0. When there is overlap between home ranges, the index can be interpreted as the average number of home ranges that overlap at any point. Calculation of the index requires a knowledge only of the overall population density of the species and the home range area of one or more troops, but the assumption is made that the habitat is uniformly occupied (see footnotes f and g in Table 7.1).

One might suppose that the amount of home range overlap would depend on the degree of tolerance shown by neighboring troops for one another. While this may be true as a rough generalization, it did not hold in every case. *Saimiri*, which shows the greatest overlap (possibly underestimated if the home range is much more than 250 ha), was also the most tolerant. There is no evident aggression when troops meet, and intermingling occurs freely and frequently, both in fruit trees and during insect foraging. *Cebus* are less tolerant of one another. This is especially true of *albifrons*, which is the exception to the pattern of increasing tolerance with increasing overlap. Troops seldom approach within 100 m of one another except unwittingly. A chance encounter results in a raucous and sometimes prolonged vocal exchange between the dominant males, and the hasty retreat of one or both groups. This behavior is strikingly paralleled in the mangabey (*Cercocebus albigena*), a species which also resembles *albifrons* in having an unusually large home range (Waser 1975a, 1975b, 1976, 1977).

Cebus apella is more variable in its reactions, but is always more tolerant than *albifrons*. Two troops may feed together in a large fig tree without incident. On other occasions there is mild repulsion, accompanied by low-level excitement and confrontations between the dominant males. Overtly hostile interactions occurred only to establish control over highly desired and spatially restricted resources, such as the nut cluster of a *Scheelea* palm. On these occasions, one of the troops was driven away from the resource, if only temporarily.

The tamarins represent the extreme of intolerance. Their strong territoriality is reinforced by frequent clashes at the boundaries. Territorial limits are sharply defined and seldom transgressed, even when no opposing group is in the vicinity. While an appreciation of these behavioral differences is important background to the present chapter, further discussion of their adaptive basis is deferred to Chapter 10.

Between Species Comparisons

Typical home range use patterns for four of the species are illustrated in Figure 7.1. *Cebus apella* and the two tamarins visit the central parts of their home ranges frequently, and only occasionally visit the perimeters. The pattern shown by *Cebus albifrons* is distinctly different in that the usage is concentrated in widely scattered foci. The maximum linear dimension is also considerably greater than those of the other species. A map for *Saimiri* is not included because it spends most of its time in mixed troops with either *Cebus apella* or *C. albifrons*. Its ranging behavior is thus linked to that of the species with which it is associated. (Further details are presented in the next chapter.)

Let us first consider the pattern manifested by *Cebus apella* and the two tamarins. Of the *a priori* models presented earlier, it conforms most closely to the expectations of central place foraging. However, none of these species uses a traditional night roost, so the main presumption of central place foraging is not in force. Instead, it appears that the pattern relates to optimal foraging in a rather simple way. Fruit resources come in discrete packets (trees) that are fixed in space. At times when suitable trees are relatively plentiful, an animal's complete home range will contain a large excess beyond the number required to satisfy nutritional needs. Under such conditions an animal or troop can economize on time and energy invested in travel by remaining within the portion of its home range that offers the greatest concentration of resources. If resources are both plentiful and well distributed, then the best place to stay

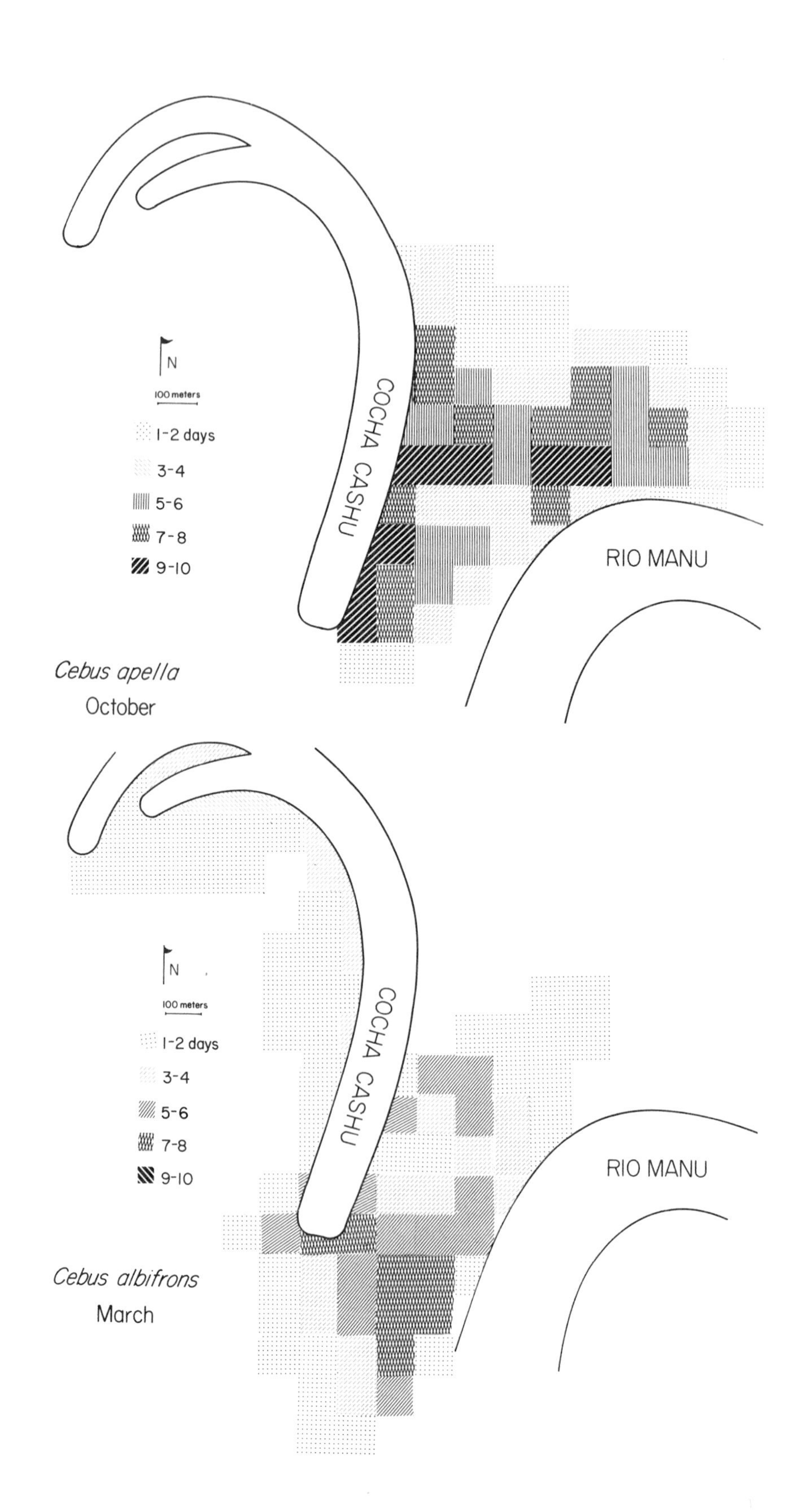
N
100 meters
1-2 days
3-4
5-6
7-8
9-10
COCHA CASHU
RIO MANU
Cebus apella
October
N
100 meters
1-2 days
3-4
5-6
7-8
9-10
COCHA CASHU
RIO MANU
Cebus albifrons
March

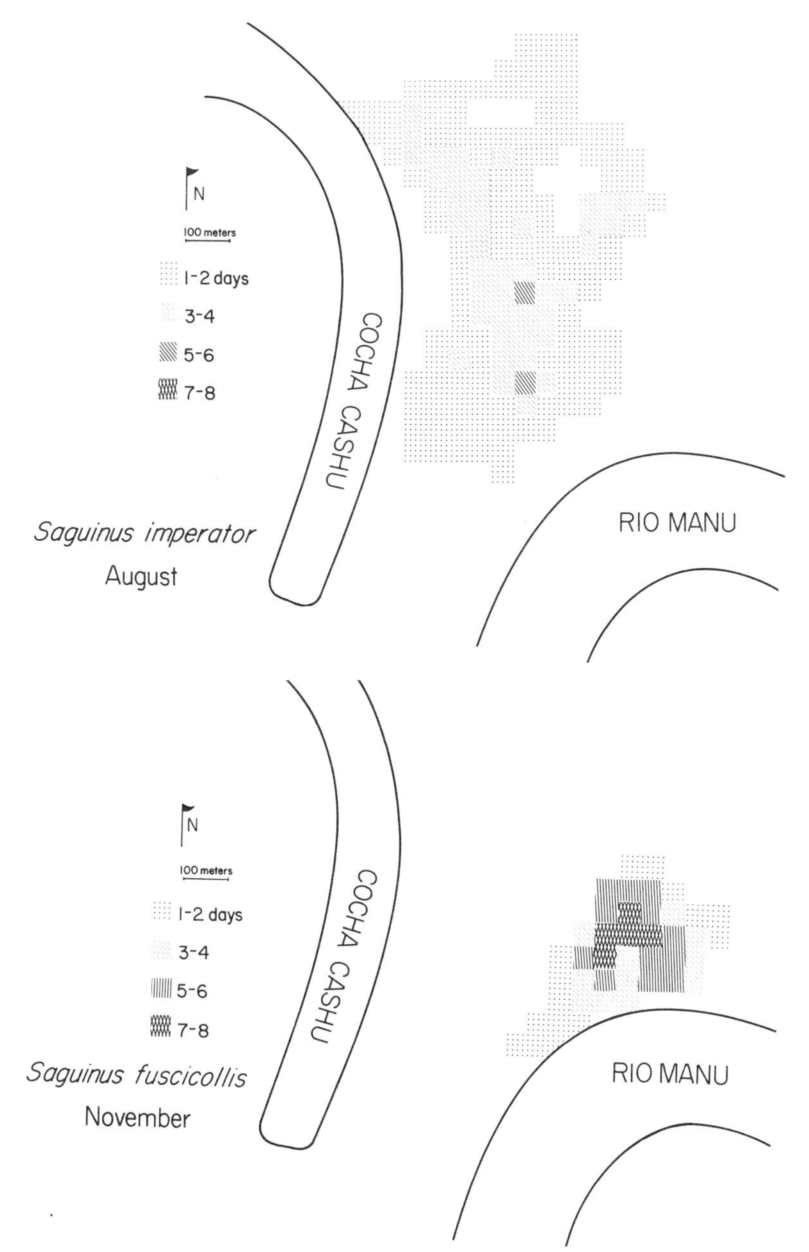

Figure 7.1 Typical home range patterns for *Cebus apella*, *C. albifrons*, *Saguinus imperator*, and *S. fuscicollis*. The degrees of shading indicate the number of days in the sample in which the troop entered a square. Grid size for *Cebus* = 1 ha, for *Saguinus* = 0.25 ha. The *Cebus apella* sample was truncated to 15 contact days to match the length of the *C. albifrons* sample. Similarly, the *Saguinus imperator* sample was truncated to 8 days to match the *S. fuscicollis* sample.

is in the central portion of the home range, because occasional reconnaissance visits to various sectors of the periphery can be made most efficiently from this location.

The situation of *Cebus albifrons* is quite different. Its home range is substantially larger than that of the other species, and is consequently that much more difficult to patrol with any regularity. During the 24-day period covered by the March sample, the troop harvested a succession of patchy fruit concentrations. The first of these was contained in a narrow strip along the western margin of the lake. Several common species were exploited as encountered in daily marches up and down the lakefront. These were *Inga marginata, I. mathewsiana* (Leguminosae), *Sloanea* spp. (Elaeocarpaceae), *Theobroma cacao* (Sterculiaceae), *Xylopia* spp. (Annonaceae) and *Casearia decandra* (Flacourtiaceae). By March 16, the last day of these treks, the stores of fruit of all these species had been conspicuously depleted. The troop then moved farther south into the flooded successional vegetation fronting the river where it harvested an abundant supply of *Cecropia leucophaia* (Moraceae) catkins and *Casearia decandra* fruits between March 17 and 27. At this juncture the troop abandoned the riparian zone and embarked on a long (>2 km) and notably unproductive two-day journey back along the familiar lakefront route. On March 28 it disappeared beyond the limits of the trail system in a large area of deeply flooded successional vegetation that occupies a prolongation of the lake bed. Presumably another concentration of fruiting trees occupied the group for the next several days because it did not reappear until April 4.

The practice of exploiting one high-density patch of resources to exhaustion, and then moving, sometimes great distances, to another is characteristic of *Cebus albifrons*. It corresponds to the shifting patch model described at the beginning of the chapter. We observed this pattern to varying degrees in every sample period, and noted it as well in other *albifrons* troops that occupied the study area.

Cebus apella behaves in an entirely different fashion. Although it travels about the same distance per day as *albifrons* (Table 7.2), its movements are much more regular and are contained within a relatively compact space. The central portion of its home range is well defined and is visited frequently regardless of the location of temporary resource concentrations. This contrast in the spatial utilization patterns of the two species is brought out in Figure 7.2. In 15 contact days *C. albifrons* visited 33% more hectares than *C. apella* (99 vs. 74), and visited each one less frequently (median 2 vs. 5 visits). Nearly half of the hectares entered by *albifrons* were visited only once or twice, and none were visited more than eight times. Of the hectares transited by *C. apella*, in contrast, 76% were visited three or more times, and nearly a quarter (24%) were visited ten or more times. Hectares in the central portion of the home range were visited every two or three days. *C. albifrons*, in contrast, may not visit major sectors of its home range for weeks at a stretch.

Table 7.2
Daily travel distances of Cocha Cashu primates.

Species	*Sample*	*No. complete days*[a]	*Mean distance per day (M)*	*Std. dev.*
Cebus apella	August	7	2,620	780
Cebus apella	October	18	2,230	420
Cebus apella	January	14	1,630	500
Cebus apella	June	17	1,820	480
Cebus apella	Mean[b]		**2,070**	**540**
Cebus albifrons	November	14	2,170	400
Cebus albifrons	March	9	1,820	400
Cebus albifrons	June	13	1,460	760
Cebus albifrons	Mean[b]		**1,820**	**520**
Saguinus imperator	August	6	1,630	230
Saguinus imperator	April	8	1,090	210
Saguinus imperator	July	3	1,540	620
Saguinus imperator	Mean[b]		**1,420**	**350**
Saguinus fuscicollis	November	7	1,160	390
Saguinus fuscicollis	April	8	1,220	450
Saguinus fuscicollis	July	4	1,290	270
Saguinus fuscicollis	Mean[b]		**1,220**	**370**
Callicebus moloch	May	4	**960**	**210**

[a] Number of days in the sample in which the monkeys were not lost.
[b] Mean of sample values.

I believe that these differences in ranging behavior are crucial to understanding the ecological distinctions between the two species of *Cebus*. With respect to the overall composition of their diets, the variety of techniques used in insect foraging, and their time budgets, they are extremely similar. Even after we had been observing them for many months, we remained puzzled about their ecological differences. It was not until the March sample just described above that we fully grasped the importance of patchy resources to *albifrons*. *Apella* also exploits temporary resource concentrations of all of the same kinds that are used by *albifrons*, but it does so only when they are available within the confines of its relatively compact home range. *Albifrons*, however, makes a specialty of using such resources, and for this reason it must integrate the landscape on a much larger scale.

Saimiri appears to go even one step further in its quest for highly concentrated resources. The epitome of such resources are figs, and *Saimiri* is a fig eater *par excellence*, as we saw in Chapter 5. The ripening of a large tree

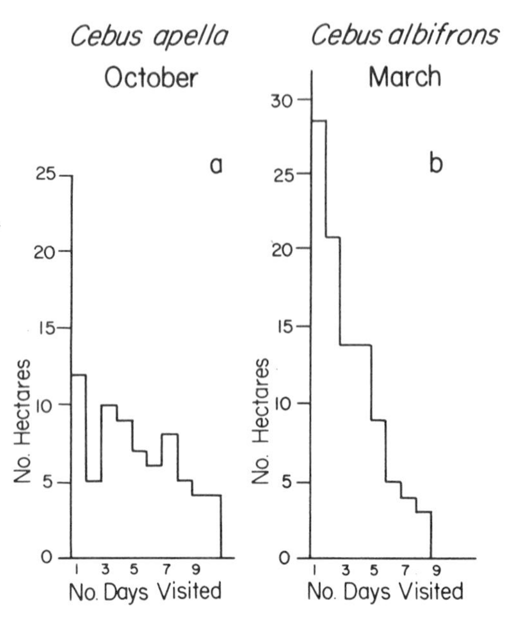

Figure 7.2 Visitation frequency histograms for *Cebus apella* and *C. albifrons*. Grid size = 1 ha. Same sample periods as in Fig. 7.1. The distributions are significantly different by chi-square test ($p < 0.001$).

will attract up to four troops at once. Normally no more than two troops use a tree at the same time, but in the course of a day all have an opportunity to feed. The animals tend to remain in the vicinity for as long as the fruit lasts, which can be up to ten days. This creates a situation that closely mimics the conditions of central place foraging (Fig. 7.3). Troops usually arrive early in the morning, gorge themselves for an hour or two, and then depart for insect foraging. They may return later in the day for a refill, or wait until the next morning. The daily foraging excursions radiate out in various directions and return via different routes. Normally the forays made on successive days cover new territory, though the immediate vicinity of the fig may be searched repeatedly.

The bustling activity at a large fig is one of the unforgettable spectacles of the forest. Monkey troops arrive from all directions as if guided by some mysterious perception. We have seen over 100 monkeys of five species and 20 to 30 species of birds feeding simultaneously in a single *Ficus perforata*. How is it possible that so many animals independently discover a tree almost the first day the fruit ripens? We believe they are summoned by the shrill din of myriads of parakeets (*Brotogerus* spp.) that quickly converge on the scene. These birds are fig specialists as much as *Saimiri*, and the sound of them in numbers is an almost certain indication of a fruiting tree.

In their obsession for figs, *Saimiri* are even more nomadic than *Cebus*

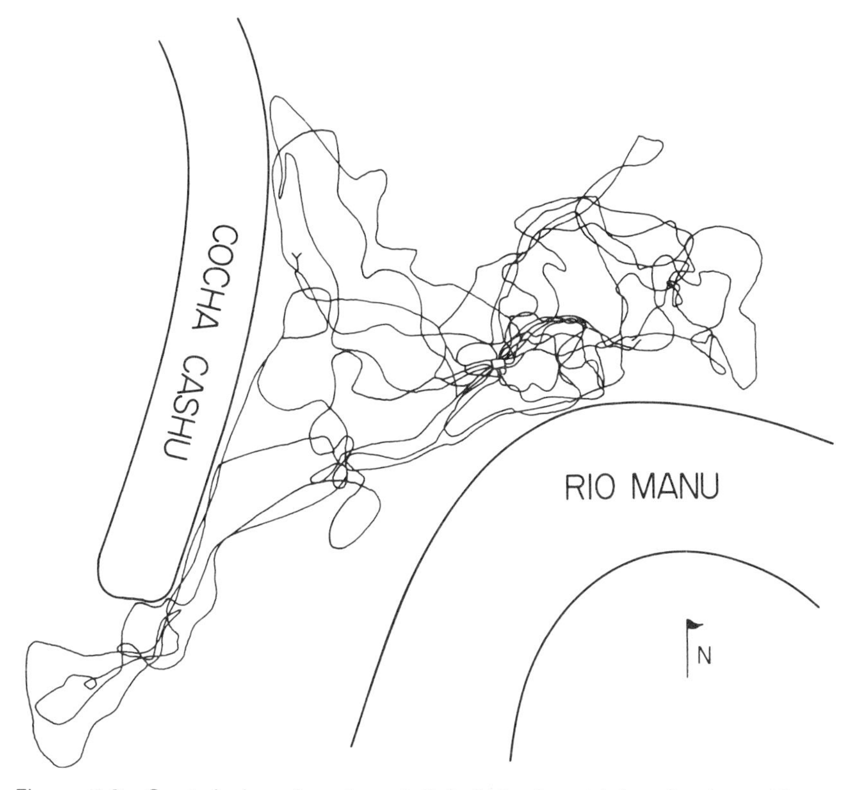

Figure 7.3 Central place foraging of *Saimiri* in the vicinity of a large *Ficus perforata*, Oct. 2–8, incl., 1976.

albifrons. After a tree has been exhausted, a troop may undertake a long odyssey in search of another. These treks may last several days and cover many kilometers. On occasion we have picked up *Saimiri* troops at one extremity of the study area and followed them until they went far out of bounds at the opposite extremity. For this reason we are unable to say just how large a home range is. We know it is larger than the 2.5–3 km^2 contained within the trail system, but how much larger is still uncertain.

Comparison of Ranging Patterns During Periods of Abundance and Scarcity

In this section we compare the ranging behavior of the study troops during the periods of maximum (November–January) and minimum (May–July) fruit abundance.

Cebus apella

The differences between the January and June samples are profound (Fig. 7.4a and b). In January (21 days) the troop used an area of about 34 ha and visited more than 225 fruiting trees for a total of 2,350 minutes. In June (20 days) the troop covered more than twice as much area (73 ha) but used only 50 trees (exclusive of palms) for a total of 890 minutes (Table 7.3). The shortfall of feeding time in June was compensated by 1,300 minutes of feeding on palm nuts and 250 minutes on pith. Thus, as stressed in Chapter 5, *Cebus apella* facultatively shifts its diet when fruit is in short supply. This is accompanied by an expansion of the area searched, but not by a change in the basic pattern of coverage.

Large numbers of fruiting *Astrocaryum* palms were available to the animals in June within the central 20–30 ha of their home range. Nevertheless, many of these were bypassed or underutilized while the troop made repeated forays into peripheral areas that were not visited in January. This can be understood if we presume that the animals prefer soft fruit and *Scheelea* nuts to *Astrocaryum*. These more desirable resources are rare and widely scattered in June. An indication of this is found in the distances between most-used trees. In January six trees accounted for 50% of the feeding time, and the average distance between all possible pairs was 330 m. Of all the soft (nonpalm) fruit eaten in June, five trees accounted for 50% of the time, and these were a mean of 540 m apart. Thus one can infer, albeit by somewhat circular logic, that the *apella* had to travel farther to discover suitable trees in June. Moreover, of the trees accounting for 50% of feeding time on soft fruit, those exploited in the June sample had less to offer than the ones used in January, as judged both by the mean number of visits per tree (4.3 in June vs. 7.3 in January) and by the mean crown diameters of the trees (12.7 m in June vs. 22.5 m in January). It appears that the animals are motivated by scarcity to go farther and farther afield in search of fruit. But even when *apella* does find attractive trees near the perimeter of its range, use of them does not alter the habit of returning frequently to the core area. Instead of staying in the vicinity of a concentrated resource as *albifrons* does, an *apella* troop feeds heartily and then withdraws toward home.

While *apella* is able to expand or contract the actively used portion of its home range without otherwise modifying its behavior, the frequency with which any given hectare is visited necessarily changes (Fig. 7.5a and b). The median number of visits per hectare in the 20–21-day samples dropped from 5 in January to 3 in June, and the maximum fell from 18 to 12. It should not be inferred from this, however, that the total intensity of use per unit area by all *apella* troops declined in June. Rather, if other troops in the vicinity

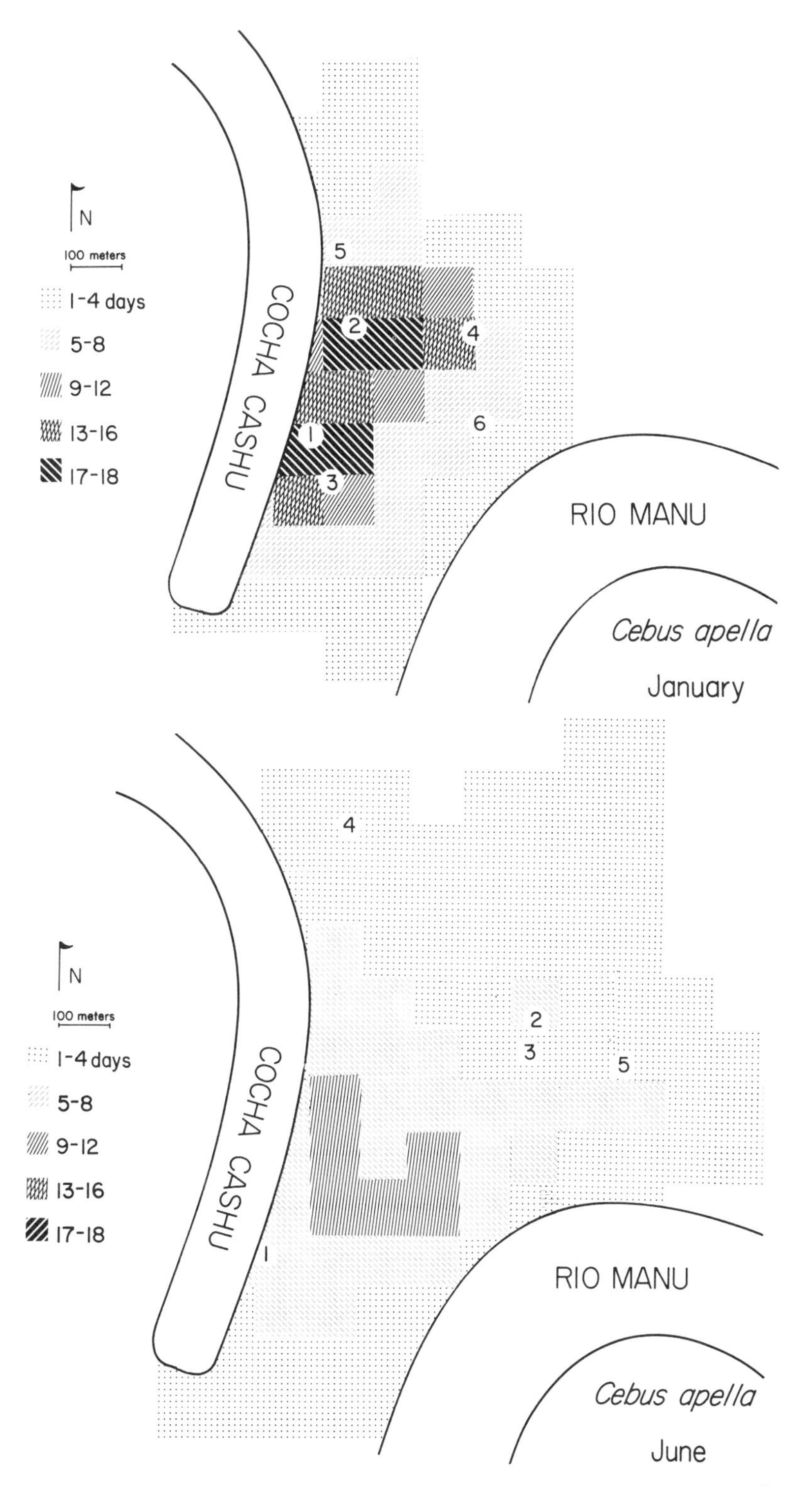

Figure 7.4 Comparison of home range use patterns of *Cebus apella* in January (above) and June (below) 1977 samples. Grid size = 1 ha. Sample lengths were 21 days for January and 20 days for June. Circled numbers indicate the locations of trees that collectively accounted for ≥50% of the feeding time for the respective samples.

Table 7.3
Use of space by *Cebus* spp. and *Saguinus* spp., by sample.

Species	*Sample*	*No. hectares used*[a]
Cebus apella	August	57
Cebus apella	October	52
Cebus apella	January	34
Cebus apella	June	73
Cebus apella	Cumulative[b]	**81**
Cebus albifrons	November	52
Cebus albifrons	March	>55
Cebus albifrons	June	>75
Cebus albifrons	Cumulative[b,c]	**≥150**
Saguinus imperator	August	30
Saguinus imperator	April	16
Saguinus imperator	July	15
Saguinus imperator	Cumulative[d]	**30,21**
Saguinus fuscicollis	November	7
Saguinus fuscicollis	April	17
Saguinus fuscicollis	July	15
Saguinus fuscicollis	Cumulative[d]	**30,24**

[a] Area within the envelope of extreme points reached, measured with a planimeter on scale maps.
[b] Cumulative area covered in all samples.
[c] Values for March, June, and the cumulative total are low due to fact that the troop was lost beyond the trail system for one or more days in these samples.
[d] Values, respectively, for the territories occupied in August 1975 and in April–July 1977.

similarly expanded their search areas, the total density of use would have remained constant while the extent of home range overlap between troops greatly increased. Indeed, this was apparent in the much higher incidence of intertroop encounters in June. Neighboring troops thus appear not to be inhibited from making deep incursions into each other's core areas, but do so mainly when resources are scarce at home.

Cebus albifrons

Here we compare samples done in November–December and June, and again major differences strike the eye (Fig. 7.6a and b). The use of space by *albifrons* during the November peak of fruit abundance was similar to that shown by

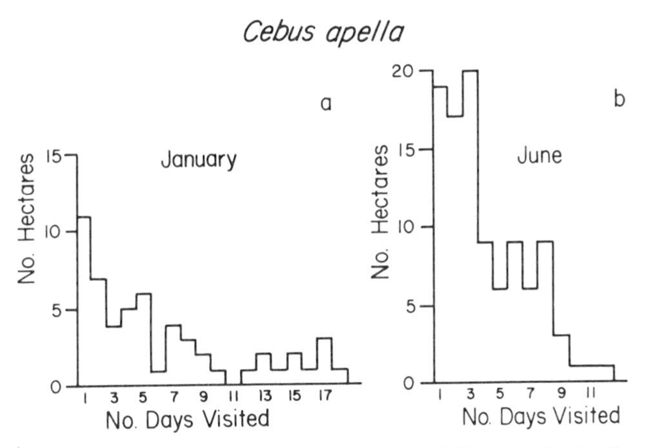

Figure 7.5 Visitation frequency histograms for *Cebus apella* in the January (a) and June (b) 1977 samples. Grid size = 1 ha. Same sample periods as in Fig. 7.4. The distributions are significantly different by chi–square test ($p < 0.001$).

apella in June. A total of 80 hectares was visited a median of three times each in 18 days (Fig. 7.7a and b). The June sample, however, represents an extreme of nomadic wandering and use of patchy resources. Half the 107 hectares entered were visited only once, and only one hectare (which contained a giant fig) was visited more than seven times.

Ripe fruit was plentiful in November. During the 18 days of the sample, the study troop fed in over 190 trees of 29 species. Many of these were in the small to medium size range (mean crown diameter of the most used ten trees that accounted for 50% of feeding time was 15.7 m). The use map shows clearly that the most frequently visited squares were the ones that contained these ten trees, demonstrating that movements were directed primarily toward the end of harvesting fruit, rather than toward other possible objectives such as insect foraging.

This inference is even better supported by the June sample in which the patchiness of spatial utilization is far more pronounced. During the 16 days of the sample, the troop fed in a total of only 19 fruit trees, and just two of these, both figs, accounted for 94% of the time spent feeding on soft fruit (Fig. 7.6b). By the overland route that the monkeys were obliged to take, the distance between these two trees was 2.1 km.

If we compare the parameters of home range and fruit tree use of *apella* and *albifrons* in the rainy season samples, the differences are rather small. Both used relatively compact areas, and both fed in a large number of trees

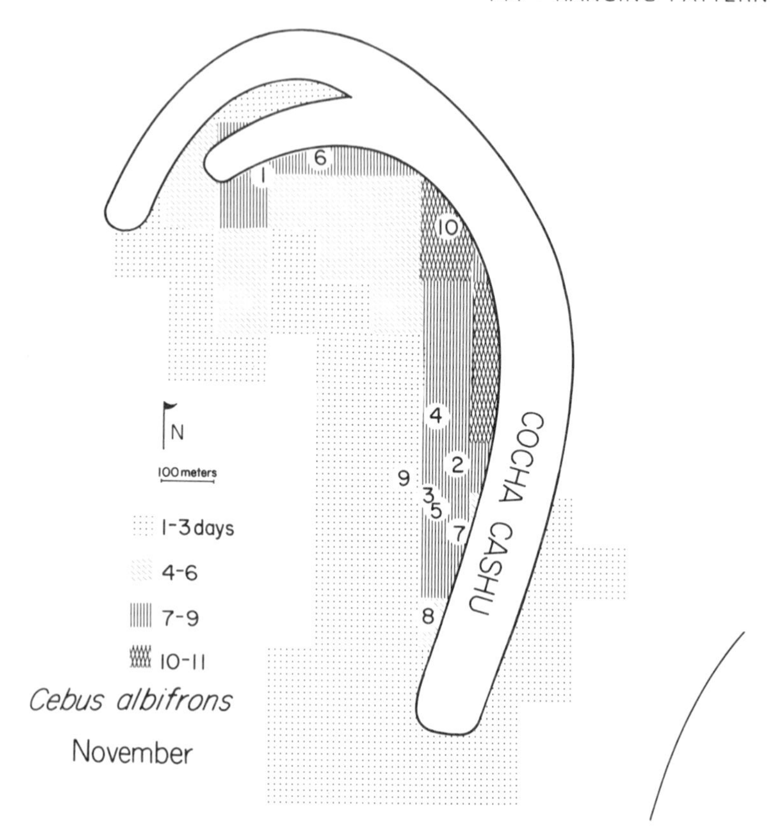

Figure 7.6 Comparison of home range use patterns of *Cebus albifrons* in November–December 1976 (left) and June 1977 (right) samples. Grid size = 1 ha. The November sample was truncated to 16 contact days to match the June sample. Circled numbers indicate the locations of trees that collectively accounted for ≥50% of the feeding time for the respective samples.

(Table 7.4). Feeding time per day was somewhat greater for *apella*, as was the mean crown diameter of most-used trees. Three of the four measures were more divergent in the simultaneous June samples. *Apella* doubled the amount of space covered while the center of its activities remained unchanged, while *albifrons* used 60% more space than in November but shifted its coverage to entirely new areas near the heart of another troop's home range. And while *apella* showed a marked decrease in fruit feeding time in June, and a decrease in the size of most-used trees, *albifrons* showed no reduction in feeding time and a major increase in the size of the most-used trees. Thus, by several criteria, the ranging behavior of the two *Cebus* species was less similar during the period of scarcity. Nearly all of these contrasts can be accounted for by

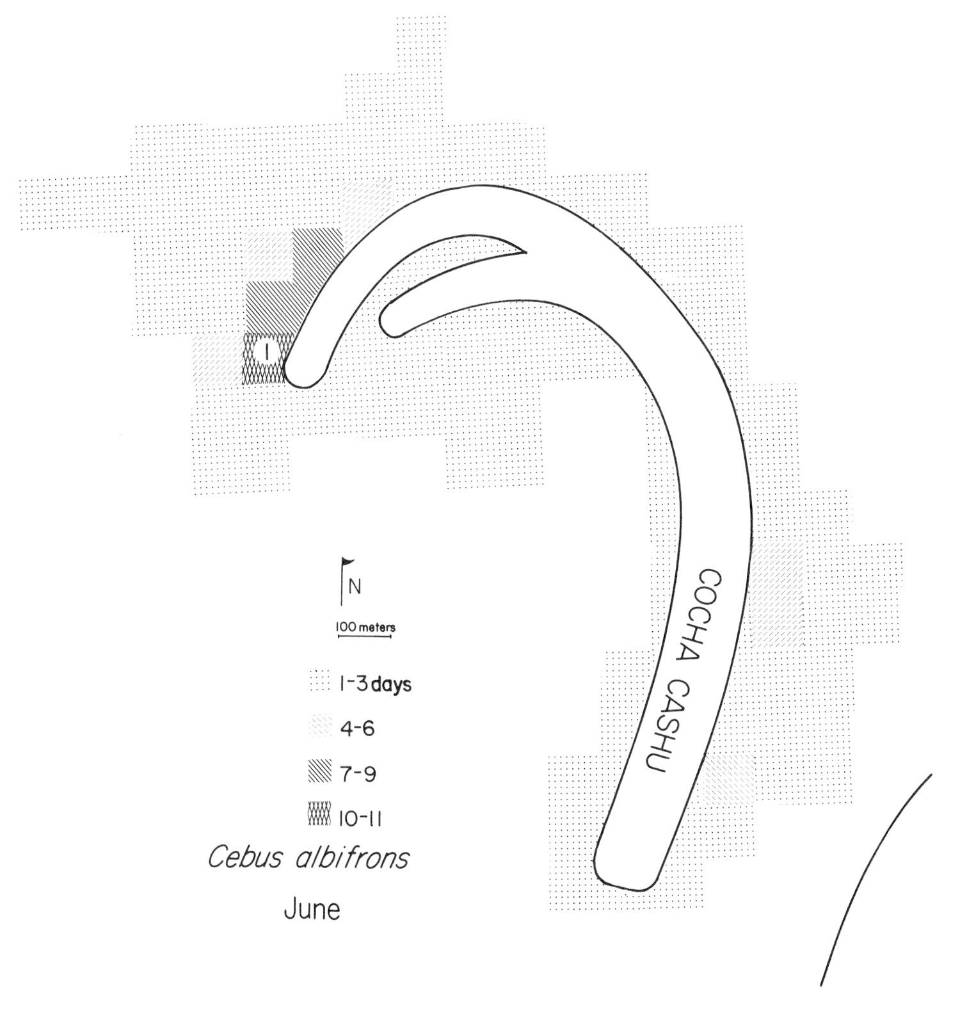

the strong dependency on figs that *albifrons* shows at this time. Attributable to this one factor are the more linear (vs. coherent) pattern of area coverage, the strong reduction in number of trees exploited, and the major increase in mean crown diameter of trees exploited.

Saguinus

Since the two *Saguinus* species live in permanent mixed associations, they shall be discussed together. Recall that the associated groups travel parallel routes and share a common pool of resource trees within a jointly defended territory. The very fact that the animals are territorial suggests that seasonal

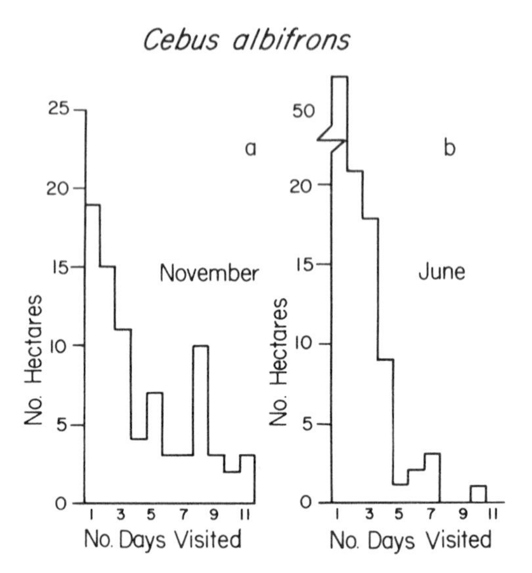

Figure 7.7 Visitation frequency histograms for *Cebus albifrons* in the November–December 1976 (a) and June 1977 (b) samples. Grid size = 1 ha. Same sample periods as in Fig. 7.6. The distributions are significantly different by chi-square test ($p < 0.001$).

Table 7.4
Comparison of home range and fruit tree use by *Cebus apella* and *C. albifrons* in samples taken during periods of maximum and minimum fruit abundance.

	Maximum Fruit Abundance (November–January)		*Minimum Fruit Abundance (June)*	
Parameter	C. apella	C. albifrons	C. apella	C. albifrons
Area covered (ha)	34	52	73 (2.1)[a]	75(1.6)
No trees used	225	190	50 (0.2)	19(0.1)
Feeding time/ day (min.)	115	86	45 (0.4)	111(1.3)
Mean crown diameter (m)	22.5	15.7	12.7(0.6)	30(1.9)

NOTE: Data refer to fruit trees exclusive of palms.
[a] Factor by which the June values are larger or smaller than the corresponding November–January values.

variations in the use of space can be expected to be less marked than in the nonterritorial *Cebus* species. This could also have been anticipated from evidence presented in Chapter 5 which showed that the long fruiting periods of major resource species probably buffer against fluctuations in the food supply. This was further supported by the fact that the number of resource trees visited per day and per sample did not vary conspicuously between the seasons. But all this is circumstantial evidence. The direct evidence currently available is unfortunately of little help. This is because one of the study troops changed either its territory or its membership between each pair of samples. These changes are described in the legend of Figure 7.8.

Given the complex history of reshuffling of groups between territories, and of individuals between groups, it is surprising that there were no accompanying changes in the locations of the territories themselves. Yet to the best of our knowledge, the boundaries between territories were in the same locations in 1979 as they had been in 1975.

The ranging patterns observed in all the *Saguinus* samples were similar in one respect. This was the pronounced tendency to use the centers intensively and the margins lightly, just as we earlier saw in *Cebus apella* (Fig. 7.9). Notwithstanding the vigor with which territorial boundaries are advertised and defended, peripheral portions of the territories are subject to little regular use. This is of course contrary to what would be expected if territorial defense were the primary determinant of movement patterns. Instead we can conclude that the animals endeavor to minimize their daily travel distance by remaining within as compact an area as possible.

Such a tendency was especially apparent in the abbreviated October *imperator* sample and in the November *fuscicollis* sample. The study troops remained within areas of five and six ha, respectively, and visited a smaller number of trees than in the other samples. *Cebus apella* similarly reduced its range when fruit became abundant, but greatly increased the number of trees used. The *Saguinus* could likewise have increased both the number and variety of trees exploited because suitable resource trees were abundant within their territories, but they did not. The reason for this conservatism remains obscure.

During the October observations, the female *imperator* was in the late stages of pregnancy and traveling with obvious difficulty and reluctance. While she often led progressions at other seasons, when pregnant she lagged behind, frequently obliging the rest of the family to wait while she caught up. By the time we turned our attention to the *fuscicollis* in early November, the female had already produced twins. The weight of the litter in tamarins is exceptionally large, being equal to 15–20% of the weight of the mother (Kleiman 1977). We can thus presume that the babies represent a substantial burden to the individual that carries them. Nevertheless, there is no indication that the troop reduced its daily travel distance at this time (Table 7.2).

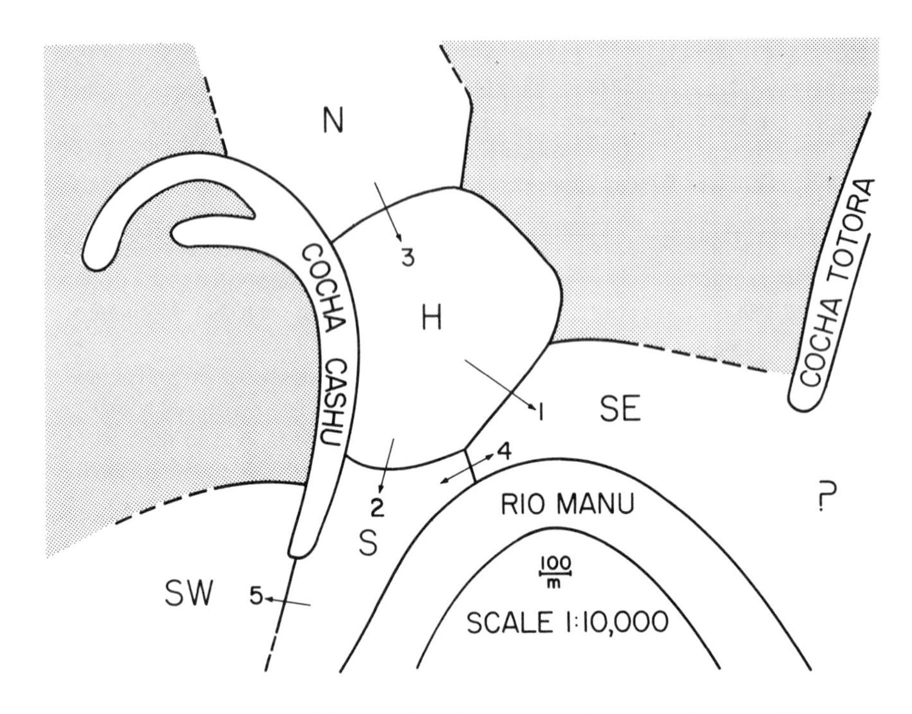

Figure 7.8 Territorial shifts by *Saguinus* troops between August 1975 and July 1977. The habituated study troops of *S. imperator* and *S. fuscicollis* jointly occupied the H territory from August 1975 until some time in September 1976. In October 1976 the *imperator* were found to be living alone in a much reduced portion of the original territory. A month later (November 1976) we relocated the habituated *fuscicollis* group in the SE territory and studied them there (arrow 1). They were sharing the territory with the intractably wild SE *imperator* troop. Late in 1976 or early in 1977 the *imperator* study troop moved to the apparently vacant S territory (arrow 2) as a large mixed group moved in from the north to occupy the H territory (arrow 3) in addition to its original N territory. Then, in April 1977, when we again studied the *Saguinus* we found that the habituated *fuscicollis* troop was alternating every few days between the S and SE territories (arrow 4). Finally, between the April and July samples the adult and subadult males of the *imperator* study troop vanished from S territory and were replaced by an unfamiliar adult male. A year later in July 1978, the missing males were rediscovered in the SW territory (arrow 5). Shading indicates portions of the study area that are not used by *Saguinus*. The various territory designations, H, N, SE, etc., refer to the locations of known *Saguinus* groups in the study area.

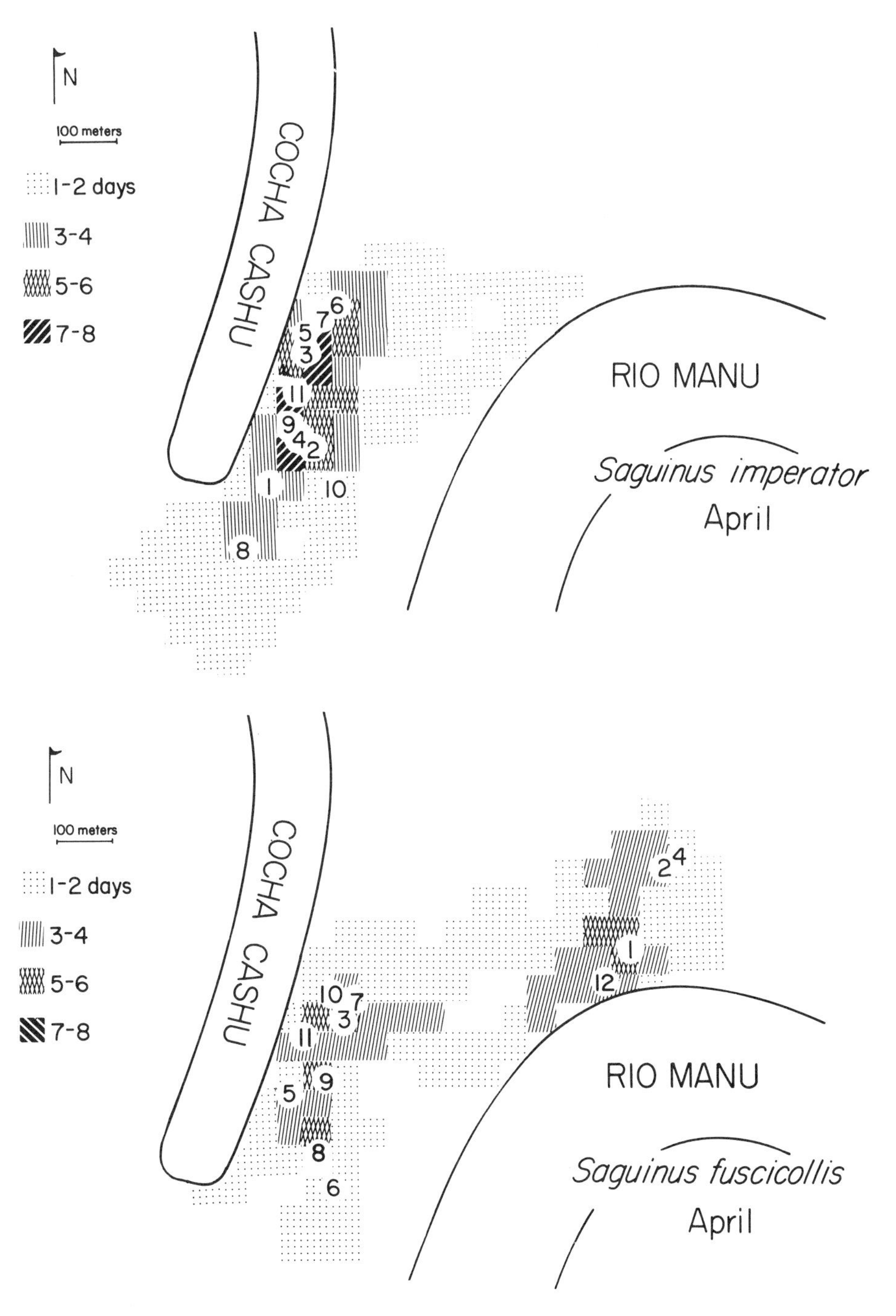

Figure 7.9 Use of space by *Saguinus imperator* (above) and *S. fuscicollis* (below) during the April 1977 samples. Intensive use of the central part of the territory is characteristic of both species. The pattern shown by *S. fuscicollis* contains two centers due to the fact that during the period of the sample the study troop was associating alternately with two *S. imperator* troops, spending 3–4 days with one and then moving to the territory of the other for a similar time. This behavior persisted through the July 1977 sample. The *S. imperator* sample was truncated to 10 days to match the *S. fuscicollis* sample. Circled numbers indicate the locations of trees that collectively accounted for ≥50% of the feeding time for the respective samples.

Conclusions

The results of this section point up some major differences between the tamarins and the larger Cebids in their use of space. The tamarins live in permanent, defended territories, and their pattern of use of these territories does not appear to vary seasonally in any important way. Even when presented with the opportunity of expanding into unoccupied habitat (cf. Fig. 7.8), they do not seem to do so. This constancy of behavior contrasts markedly with the pronounced seasonal shifts in feeding and ranging habits that characterize the other species. What is special about the tamarins is the nature of their plant resources. All the major resource species are common and ripen their crops a few fruits at a time over long periods. Appreciable seasonal overlap between successively used major resources assures a long-term continuity of the food supply within the bounded space of the territories (cf. Chapter 5). In contrast, continuity in space and time are conspicuously lacking in the plant resources most used by *Saimiri* and *Cebus albifrons*. *Cebus apella* falls in an intermediate position by virtue of its facultative ability to switch to palm nuts and pith when there is little fruit available. While *apella* is not territorial in the sense of defending a definite space, neither is it compelled to wander far from the familiar center of its range when the going gets rough. The considerable degree of spatial homeostasis *apella* does achieve is through its muscular strength and dietary versatility. The tamarins achieve an even greater consistency, but by the very different means of selective use of both habitat and resources.

Comparison of the Space Requirements of Old and New World Primates

Some intriguing contrasts stand out when one compares population and dietary statistics for the species included in this study with those of Old World monkeys (Table 7.5). All the Old World species include some leaves in the diet and most consume substantial quantities of them. The highest biomasses, as Hladik (1975), Clutton-Brock and Harvey (1978), and others have frequently noted, are associated with high levels of folivory. Another distinction is that the Old World species spend much less time searching for animal prey, if one takes at face value figures derived from several different sampling techniques.

How can one interpret these contrasts? This is a difficult question to answer with any confidence because the five New World species differ in at least three important respects from their Old World counterparts: they are smaller, they do not eat leaves, and they are much more insectivorous. It seems likely

Body weight, home range area, group size, population density, biomass, and percentage composition of diet of some New and Old World forest primates.

Species	Adult body weight (kg)	Home range area (ha)	Mean group size	Population per km²	Biomass (kg per km²)	Percentage of diet[a] Leaves, flowers	Percentage of diet[a] Fruits, seeds	Prey	Ref.
NEW WORLD SPECIES									
Cebus apella	3.0	80	10	40	105	0	24	76	
Cebus albifrons	2.8	≥150	15	35	85	0	36	64	This
Saimiri sciureus	0.8	>250	35	60	50	0	18	82	study
Saguinus spp.	0.35	30	4	28	10	0	42	58	
OLD WORLD SPECIES									
Cercopithecus mitis	4.5	14	14	42	190	35	38	11	
Cercopithecus cephus	3.5	35	10	25	88	8	79	10	
Cercopithecus nictitans	5.4	67	20	30	160	29	61	8	
Cercopithecus pogonias	3.7	78	15	23	85	2	84	14	Clutton-Brock
Miopithecus talapoin	1.2	120	70	40	48	4	52	43	and
Cercocebus albigena	7.7	410	15	3	21	8	61	24	Harvey
Macaca fasicularis	5.0	32	23	50	250	21	52	23	1977a
Macaca nemestrina	8.5	200	35	20	170	23	72	2	
Macaca sinica	4.6	38	13	100	460	12	85	2	
Colobus badius	8.7	67	41	185	1600	87	8	0	
Colobus guereza	9.8	15	11	104	1000	84	14	0	
Presbytis melanophos	6.1	21	9	74	450	46	53	0	
Presbytis obscurus	6.1	29	10	31	190	55	46	0	
Presbytis senex	8.2	12	8	150	1200	72	28	0	
Hylobates lar	5.5	54	4	5	26	31	59	10	
Symphalangus synd.	10.7	23	4	4	45	49	42	8	

[a] All figures for dietary proportions are based on time spent, not weight or volume of food eaten.

that the high degree of insectivory is made possible by the small size of the New World species, but it does not necessarily explain their size. Neither does their size provide a convincing argument for the complete lack of folivory, because there are a number of folivorous lemurs within the same range of body weights (Hladik 1979). Among both New and Old World species, however, it seems clear that leaves substitute for animal prey as body size increases. This is seen in the partial folivory of howler and woolly monkeys in the New World and is especially apparent in the contrast between the small 1.2 kg *Miopithecus talapoin* and most other Old World monkeys. Nevertheless, in comparing the two groups of species, there remain riddles we cannot solve, such as the fundamental contrasts in body size and prevalence of folivory. I will come to these issues again in Chapter 9.

Another distinctive shared trait of the New World species is that their home ranges are especially large in relation to the biomass per troop. Clutton-Brock and Harvey (1977b) have shown that frugivores have larger home ranges in general for a given group weight than folivores. But comparing our results to theirs, we find that all five species lie well above the regression line for frugivores; that is, their home ranges are larger than is typical for a wide selection of mostly Old World genera. Three possible explanations come to mind. It could be that the trend line is depressed by the partial folivory of many of the taxa included in the regression (e.g., *Cercopithecus, Macaca, Hylobates*, etc.). Another possibility, one which cannot be tested at present for lack of appropriate data, is that fruit is less abundant and/or more highly seasonal in New World forests. A third possibility is that insectivory imposes an especially large requirement for space, as suggested for *Macaca sinica* by Hladik (1975).

There is nothing in our observations to lend support to this idea, and a good deal of evidence to cast doubt on it. The large home ranges of *Cebus* and *Saimiri* are due to the broad overlap of areas used by neighboring troops. It is hard to see why such overlap would occur if arthropods were the controlling factor unless arthropods were highly clumped in the environment. The slow, steady progress of foraging troops, however, implies that the dispersion of prey is rather uniform. A counterexample can be made of insectivorous birds living in forested habitats. These are almost universally territorial (Crook 1965). Moreover, much of the evidence presented in this chapter suggests that the patterns of movement and spatial utilization of the five species are controlled by their need to obtain fruit. This includes the nomadism of *Saimiri*, the concerted use of temporary fruit concentrations by *Cebus albifrons*, the expansion of the home range of *Cebus apella* during the season of fruit scarcity, the nonuniform spatial utilization patterns of all the species, and the fact that all of them do much of their foraging for prey while moving between fruit trees. None of this is direct evidence, but as circumstantial evidence the cumulative effect is persuasive. We must therefore turn to one or both of the other possibilities in seeking an explanation for the large home ranges of our species. It is very likely that by including partially folivorous species, the trend line computed by Clutton-Brock and Harvey (1977a) is too low. It is also possible that the long period of fruit scarcity that compels *Cebus* and *Saimiri* to expand their ranges is more extreme than in many Old World tropical localities. A limited amount of data on fruit phenology in the Malayan forest suggest that this might be the case (Raemaekers et al. 1980).

Summary

Parameters of the ranging behavior of the five species are compared *inter se*, and then the behavior of each species is compared with itself during the periods of maximal and minimal fruit abundance. In spite of very similar dietary requirements, the species show great differences in home range size, home range overlap, expressions of territoriality, and concentration of spatial utilization within home ranges. The behavior of *Saimiri* lies at one extreme. In its seminomadic use of an extremely large home range (>250 ha) and in its tolerance of conspecifics, the species is adapted to harvesting concentrated but spatially and temporally unpredictable resources, of which figs are the prime example. The tamarins lie at the opposite extreme. They live within rigidly bounded territories and feed on a limited number of common resource species that ripen crops over long periods. These characteristics of tamarin resources assure a long-term continuity of supply in both space and time. Because of this, the tamarins receive an advantage in defending a delimited space for their exclusive use.

The two *Cebus* species represent intermediate conditions. *Cebus albifrons* is more like *Saimiri* in its dependence on shifting resource concentrations and seasonal habitats. The home range is large (ca. 150 ha), and use of it within any few-week period is extremely uneven. Even over a year, there are only weak indications of a core area. *Cebus apella* is less dependent on soft fruit than *albifrons*. In the dry season it can switch to abundant but less preferred resources such as palm nuts and pith. Along with the tamarins, it ranges in an intensively used core area encased within a lightly covered peripheral zone. It differs from the tamarins, however, in greatly expanding the peripheral zone during the season of fruit scarcity. But unlike *albifrons, apella* uses the central part of its home range consistently at all seasons.

The feeding and ranging behaviors of the two *Cebus* species are highly similar when fruit is superabundant, but pronounced differences develop during the period of scarcity. The species diverge in the amount of space covered per sample, the amount of time spent feeding on soft fruit, the number of trees visited, and the mean size of the trees used. In contrast, the tamarins, with their more buffered food supply, show little indication of seasonal changes in their ranging patterns.

At the end of the chapter, I pointed out some ways in which the five species collectively differ from most Old World primates: they are conspicuously smaller in body size, they do not consume foliage, they are far more insectivorous, and they have exceptionally large home ranges per unit of biomass. A great deal of circumstantial evidence indicates that the home ranges of the five species are determined by the need to maintain an adequate intake of fruit through the dry season. Although arthropods and other small animals are important in the diets, there is little to suggest that any of the ranging patterns are primarily conditioned by the search for prey. I conclude that the smaller home ranges of many larger Old World primates are encouraged by the partial to complete substitution of leaves for animal matter as the principal source of protein, and possibly by a decreased seasonality of fruit production in Old World forests.

8 Ecology of Mixed Troops

The study and understanding of mixed-species groups of primates can help us learn about the ecological benefits of sociality. First, there is usually no possible basis for social attraction between species based on access to sexual partners. Second, there is no influence of kinship or inclusive fitness to complicate social behavior between species. Third, the specificity of such associations may point to the selective forces involved (e.g., Booth 1956; Thorington 1967; Gartlan and Struhsaker 1972; Klein and Klein 1973; Gautier and Gautier-Hion 1969; Gautier-Hion and Gautier 1974). Thus mixed-species associations offer a simplified form of sociality which must almost certainly be based on advantages with respect to food finding, predator avoidance, or some other mechanism not involving sexual or genetic affinities.

The array of proposed explanations for mixed-species associations is large. It includes: (1) improved predator detection or avoidance (Gautier and Gautier-Hion 1969; Gartlan and Struhsaker 1972); (2) improved foraging efficiency by avoidance of previously used areas (Cody 1971); (3) increased insect capture rates by flushing prey or by one species ''parasitizing'' the capture ability of others (Klein and Klein 1973; Rudran 1978; Munn and Terborgh 1980); (4) sharing or parasitism of knowledge about local scarce food sources (Gartlan and Struhsaker 1972); (5) increased encounter rate with food without parallel increase in competition (Gartlan and Struhsaker 1972); and (6) increased dietary breadth by increased movement speed of associations (Gautier-Hion and Gautier 1974). It should be noted that the last explanation here does not require association as such, but only modified behavior by the monkeys; therefore, it seems an unlikely evolutionary factor in causing mixed-species groups. Furthermore, three of the remaining explanations (nos. 2, 4, and 5) involve unequal distributions of benefits to the participating species. In such cases, one of the species is being mildly parasitized by the other(s), or else the benefits exchanged are different in kind; for example, one species could gain in predator avoidance, while the other benefits from increased access to food.

When one member of an association receives no benefit, its participation should be passive; active joining, in turn, should indicate some anticipated benefit. Thus the distinction between passive and active participation can be crucial to interpreting the signs and magnitudes of exchange benefits. In any case, both the costs and benefits transferred in an association will depend heavily on the degree of ecological similarity of the participants and on the prevailing environmental circumstances (e.g., scarcity of food). Many bird species that habitually join mixed flocks at certain times of year (usually the winter), do not do so at all at other times of year (e.g., when nesting; Morse 1970). It is therefore important to be alert to seasonal variations in the frequency and duration of associations, because the circumstances of the variations can provide additional clues to the forces that motivate the behavior.

Much of the following discussion of mixed primate troops at Cocha Cashu will evolve around the two issues just mentioned: whether participation is active or passive, and the interpretation of seasonal variations in behavior. We shall see that the interactions involved in two distinct types of association are very different. I begin with brief descriptions of the associations.

Saimiri-Cebus Associations

One type of association involves *Saimiri* with one or the other species of *Cebus*. On infrequent occasions, and then only briefly, all three species may travel together for a time. Normally, the mixed groups consist of one troop of *Saimiri* with one of *Cebus*. Less commonly, just a few *Saimiri* (i.e., 1–5, typically subadult males) will travel with a *Cebus* group. Such detached cliques of *Saimiri* are virtually always in the company of other monkeys, preferentially *Cebus*, but occasionally some other species such as *Callicebus, Saguinus*, or even *Ateles*. The duration of *Saimiri-Cebus* associations is highly variable. They may persist for as little as a few hours or for as long as ten days, or perhaps even more.

During travel and foraging there is extensive intermingling of the species. Nevertheless, physical contact and agonistic interactions are infrequent. Overt antagonism is much more conspicuous in fruit trees, especially small ones. *Cebus* often allow *Saimiri* to feed in close proximity, but at times react vigorously to their presence. Most often a single *Cebus* merely supplants a single *Saimiri* from a desired feeding station, but occasionally an adult *Cebus* will lose its patience and rush pell-mell around a crown, expelling all the *Saimiri* in it. The *Saimiri*, however, are not readily intimidated and very soon begin reentering the crown on the side opposite the irate *Cebus*.

The daily activity periods of *Saimiri* and *Cebus* are very nearly the same, a fact that allows the associations to continue from day to day without interruption. At dusk the troops usually move 50 or 100 m apart to spend the night in separate roosting areas. The groups then merge again promptly and unceremoniously the following morning as they embark on the daily round of foraging.

Mixed *Saguinus* Associations

The second type of mixed association at Cocha Cashu involves the two *Saguinus* species. The usual configuration consists of one family unit of *S. imperator* and one of *S. fuscicollis*. In our experience combined group sizes have ranged from five (three *imperator* and two *fuscicollis*) to sixteen (eight of each). Mixed *Saguinus* groups often encounter *Callicebus* families, and when they do, all three species may move off together or remain in proximity for a rest period. Although there is clearly an attraction of some kind between *Callicebus* and *Saguinus*, the associations are invariably short-lived.

Unlike *Saimiri-Cebus* associations, the coalitions formed by the two *Saguinus* species are permanent, or nearly so. Two families that were living together in 1975 were still associating, though in a different territory, in 1977 when the *imperator* family disappeared. Although we are less certain of the identity of the individuals, another mixed *Saguinus* group seems to have remained intact in an adjacent territory for the three-year period from 1976 to 1979. Times as long as this may approximate the life expectancy of *Saguinus* groups. Partners in the mixed associations share and jointly defend a common territory, a most remarkable habit that is also possessed by certain tropical insectivorous birds (Munn and Terborgh 1980). Patrolling, vocal advertising, and physical defense of territorial boundaries is conducted simultaneously by both species, though each directs its attentions toward its conspecific adversaries in the neighboring mixed group.

In another departure from the *Saimiri-Cebus* model, *Saguinus* families generally do not intermingle during travel and foraging. For one thing, their radically divergent substrate preferences for insect hunting oblige them to move at different rates, in different microenvironments, and at different heights. But even when two associated families are merely traveling from one part of their territory to another, they often move along separate though parallel paths. Because of their more frequent use of cling-and-leap locomotion, the *fuscicollis* tend to spread out and to progress in a series of rushes and pauses, while the *imperator* move more steadily and deliberately in close single file along branches and vines. The pronounced differences in foraging habits and locomotory patterns preclude a tight cohesiveness of the groups. The members of each family are nearly always closer to one another than any of them is to a member of the partner group. Often the two families are out of visual contact, but this is compensated for by frequent vocal exchanges that allow coordination of movements.

Much feeding is done by *Saguinus* in small trees or vines that will not accommodate or satiate a large number of animals. Dominance interactions regulate priority of access to crowns and feeding sites. In large groups (e.g., an eight-member *fuscicollis* family) it often happens that not all members can feed at once. In these situations one or more individuals are obliged to wait at the base of the tree while the others feed. Then, when the first individuals begin to descend, the remaining group members rush up and search the crown for leftovers. This is a sequence that one sees every day in *Saguinus* troops. It may involve differential access of the members of a family to feeding sites, or sequential access of the two species. In interspecific dominance interactions, *imperator* always wins. Even a half-grown juvenile readily supplants an adult *fuscicollis*. Such incidents are usually resolved quickly and without much overt aggression or vocal protest.

As in *Saimiri-Cebus* associations, the two species seek separate overnight retreats. Each family has a series of traditional roost sites. These are rotated in an irregular schedule, but a given site is rarely used on two successive nights. *Imperator* shows a preference for isolated, vine-shrouded trees that command a good view of the surrounding forest. We have never seen them roost in stick nests or in the outer branches of open crowns, as recorded for *Saguinus oedipus geoffroyi* (Dawson 1979). *Fuscicollis* also uses exposed vine-draped crowns similar to the ones selected by *imperator*, but sometimes roosts in hollow trunks. The two species generally pick nearby sites, but sometimes diverge to locations that are as much as 100 m or more apart. Loud vocalizations given back and forth then serve to reestablish coordinated movements the following morning.

I now proceed to a detailed analysis of the two types of association.

Saimiri-Cebus Associations

Some clues to the incentives promoting these associations may be gained by determining whether the roles played by the two species are similar or different, and whether both species are equally active in their participation. Several lines of evidence may be examined to infer the answers to these questions. First, there is the issue of which species determines the movements of the mixed group; the following species can decide to remain or not with the leader, while the leader cannot force the other to follow it. Second, it may be possible to discern differences in the behavior of a species when it is and is not participating in a mixed troop. If there is an imbalance in the benefits derived by the participants, then the species deriving the greater benefit may show the more pronounced change in behavior. And third, direct observations of joining and splitting can indicate which species is the more active participant.

Leadership of Mixed Groups

It is not easy to quantify the roles of the species in the determination of group movements. The same problems that have been encountered in studying the issue of "leadership" within other primate groups apply here, especially the complex interaction between those individuals that are first in a progression, and those that decide to follow or not (e.g., Kummer 1968; Altmann and Altmann 1970). Mixed groups of *Saimiri* and *Cebus* typically include 45–55 monkeys. *Saimiri* are usually in the forefront, while the last individuals are often *Cebus* (Table 8.1). The consistency of this observation had us confused for a long time. Taken at face value, it implied that the *Saimiri* were leading, but this interpretation did not jibe with certain other observations. It turned out that the situation is a good deal more complex than it seemed.

If individuals were randomly distributed through the mixed assemblage (an unlikely possibility), one could expect a higher frequency of first-individual records for *Saimiri* simply as a consequence of the greater dispersion of the troops and of the larger number of individuals. But this simple view of the

Table 8.1
Lead species in mixed *Saimiri-Cebus* troops
in three types of progressions.

	Activity		
	Insect foraging	*Travel to fruit tree*[a]	*Rapid travel*[b]
Saimiri accompanying			
Cebus apella			
Saimiri ahead	47	7	1
C. apella ahead	17	11	11
Cebus albifrons			
Saimiri ahead	17	5	0
C. albifrons ahead	3	2	2

NOTE: Each observation represents an independent progression. Most observations were made on different days. When two observations from the same day were included, they were widely separated in time and by intervening activities such as visiting a fruit tree. The results are significantly nonrandom for *C. apella* ($p<0.01$), but nonsignificant for *C. albifrons*.

[a] Progressions that ended with the arrival of one or both troops at a fruit tree.

[b] Rapid progressions (without insect foraging) that did not end at a fruit tree.

question does not provide an explanation for the observation that the species identity of the leading individual appears to depend on the circumstances (Table 8.1). When a mixed group was insect foraging, a quiet slow-paced activity in which the animals were often widely spaced, *Saimiri* were found to be in the lead about three-quarters of the time. However, when the animals were traveling rapidly, either (as was later ascertained) to a fruit tree or to a new foraging area, *Cebus* more often led. This could be interpreted as an indication that individuals do spread out more or less randomly during foraging, while major moves are guided by *Cebus*.

On many occasions we noticed that *Saimiri* ran ahead of *Cebus*, only to stop and reverse course if the *Cebus* did not follow (Fig. 8.1). Another common situation arose when the *Cebus* stopped at a grove of palms or at some other location that was of interest only to them. When this happened the *Saimiri* either waited quietly until the *Cebus* finished, or they scattered and foraged around the periphery of the area being used by the *Cebus*. Extreme examples of this occurred when the *Cebus* (generally *C. apella*) discovered a cluster of ripe *Scheelea* (palm) nuts. Exploiting the nuts would often occupy the troop for several hours. Since the *Cebus* always dominated the tree, and since in any case the *Saimiri* were too small to handle the unopened nuts, the two species had divergent interests. The usual consequence was that the majority of the *Saimiri* departed on independent foraging excursions from

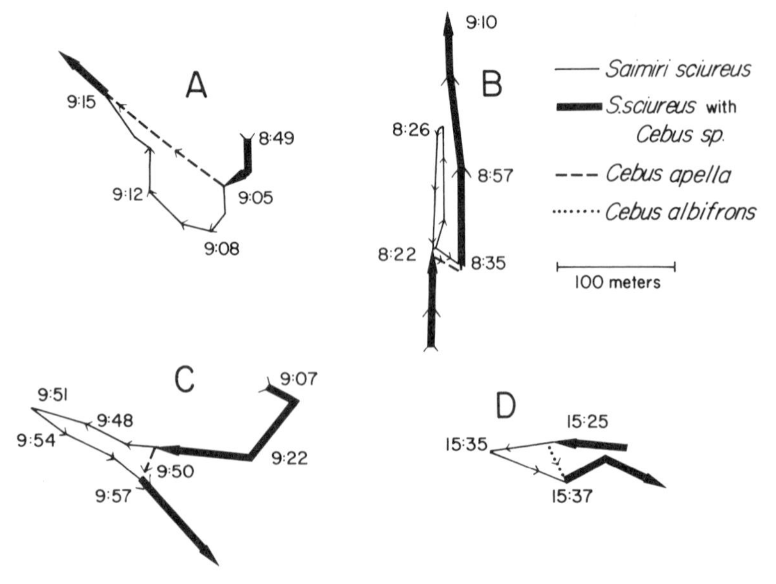

Figure 8.1 Overshoots by *Saimiri* during association with *Cebus* sp. A, B, and C with *Cebus apella*; D with *Cebus albifrons*.

which they returned every half hour or so to check on the activity of the *Cebus* (Fig. 8.2). During one of these periods the *Saimiri* sometimes made a number of closed loop forays, going out from the *Scheelea* and returning, each time taking a different route. This behavior implies that the *Cebus* constitute an attraction for the *Saimiri*, but it does not so clearly point to conclusions about the leadership of group progressions.

By contrast, it was rare to see instances in which a *Cebus* group stopped in apparent response to *Saimiri*, and in none of these few cases was it evident that the *Saimiri* had a particular reason for halting where they did. When *Saimiri* stopped to feed in trees not used by *Cebus*, the latter tended to keep moving and to outdistance the *Saimiri*, which then caught up later. This kind of observation argues that *Cebus* is the more important species in controlling group movements.

Nevertheless, we observed situations in which *Cebus* appeared to have contributed actively to maintaining an association, or conformed to a movement initiated by *Saimiri*. It was not uncommon for a *Cebus* group to move off in the direction of an associated *Saimiri* troop when the *Cebus* exited from a fruit tree in which they alone had fed. Unfortunately it is difficult to judge the motivation involved in such occurrences, because it appears that *Saimiri*

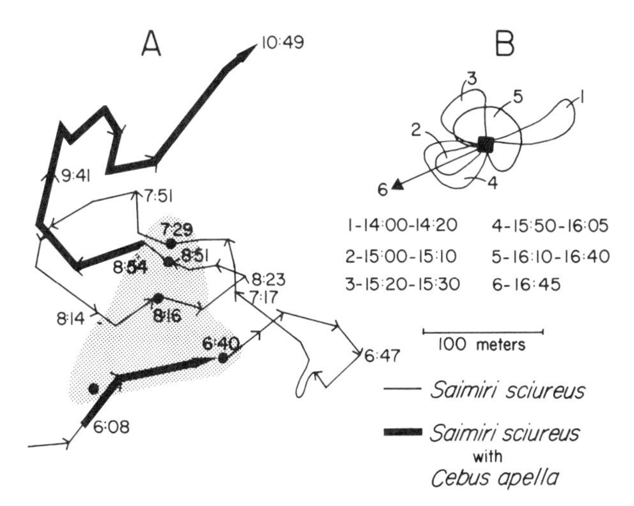

Figure 8.2 Movements of *Saimiri* during periods of palm foraging by associated *Cebus apella* troops. A: In this sequence the *Cebus* foraged for nearly 3 hours in a group of *Astrocaryum* palms (shaded) while the *Saimiri* ranged more widely in the vicinity, periodically reestablishing contact with the *Cebus* (circles). B: Here the *Cebus* fed on the nuts of a *Scheelea* (square) for almost 3 hours while the *Saimiri* made repeated insect-foraging excursions in the neighborhood.

can and do anticipate the movements of *Cebus*. Foraging *Cebus* often move for hundreds of meters along relatively straight trajectories, so that extrapolating the course may be quite easy in the short run. On at least two occasions, one with *C. apella* and one with *C. albifrons*, the accompanying *Saimiri* troop moved far ahead, traveling rapidly. In both instances the *Saimiri* ended up in a fruit tree not previously visited in the sample by the *Cebus*. Again, the possibility of anticipation clouds the interpretation of such isolated cases. Perhaps in these instances the *Cebus* did follow the *Saimiri*, but the foreknowledge of a fruit tree ahead would have provided sufficient impetus in itself. In any case, the number of instances in which *Cebus* can be argued strongly to have been following the *Saimiri* is very small relative to the number of cases in which *Saimiri* conformed to patterns set by *Cebus*.

Some striking demonstrations of the preeminent role of *Cebus* in directing movements are contained in certain instances in which *Saimiri* troops joined or transferred between *Cebus* troops. Most of the monkeys living near Cocha Cashu are not shy of humans, but a few groups that occupy the less frequented corners of the study area are still quite wild. On several occasions we were following a habituated group when it joined or was joined by a shy group of another species. In every case in which a tame *Saimiri* troop joined a shy *Cebus* troop, whether *apella* or *albifrons*, the *Saimiri* maintained the association even when the *Cebus* fled at high speed or engaged in evasive tactics such as doubling back on their course or moving in figure-eights (Fig. 8.3). In contrast, when an untame *Saimiri* group joined a habituated *Cebus* group, we did not notice any changes in the behavior of the *Cebus*. Instead, the *Saimiri* maintained a distance of 50–100 m from the observer while nevertheless continuing to follow the course set by the *Cebus*. Thus tame *Saimiri* troops may drastically modify their normal pace for the sake of maintaining associations with terrorized *Cebus* troops, but there are no indications that *Cebus* are willing to reciprocate in the converse situation. If both species were equally eager as participants, then one would not expect to see such an asymmetry of behavior.

Behaviors In and Out of Association

Saimiri-Cebus associations offer an unusual advantage to the investigator in that the interacting species can be observed both in and out of association. This opportunity is not usually available to students of mixed bird flocks because the principal species forage almost exclusively in each other's company, and because unassociated individual birds are very difficult to find, much less to follow for prolonged observation. In the present situation, measurements of the behavior of a species before and after forming an association

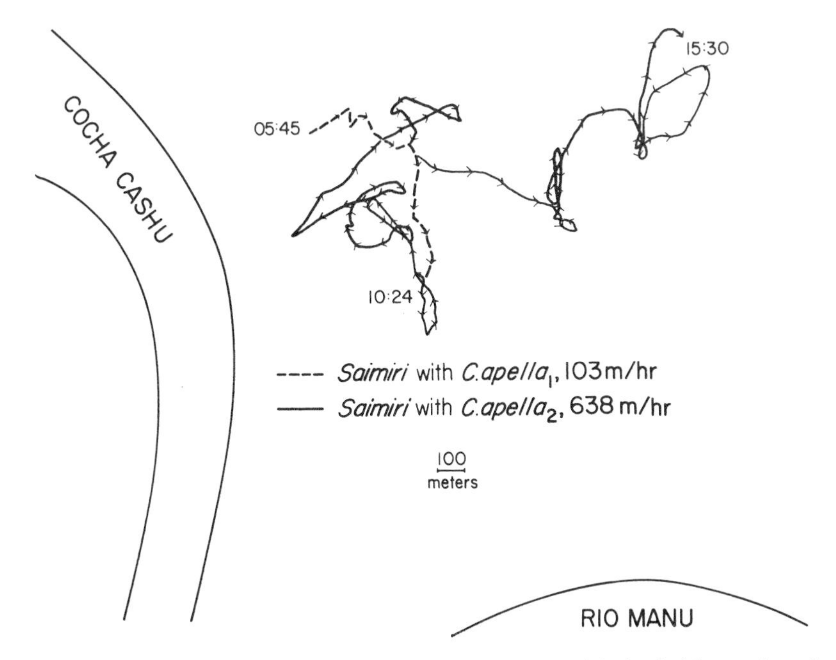

Figure 8.3 Travel of a habituated *Saimiri* troop, first with the habituated study troop of *Cebus apella* (5:45 to 10:24) and then with a wild *apella* troop. The wild troop undertook a series of switchbacks, loops, and figure-eights in an effort to evade the observer, while the panting and exhausted *Saimiri* troop doggedly followed along.

permit a variety of inferences about roles in the mixed group and about the nature of the benefits derived from the interaction.

Further evidence that *Saimiri-Cebus* associations do not affect both species equally is the much greater change seen in *Saimiri* when joining or leaving an association. Although the number of times we were following *Saimiri* troops when a *Saimiri-Cebus* association broke apart is relatively few, the behavioral changes in the *Saimiri* group once it was alone were generally striking and consistent (e.g., Fig. 8.4). The *Saimiri* usually moved far beyond the range of the *Cebus* group they were previously with, often traveling rapidly until they were a kilometer or more from the point of disassociation. If another *Cebus* group was not encountered right away, there then began a period of rest and slow foraging lasting up to several hours, followed by another rapid movement. This stop-go behavior continued until another *Cebus* group was encountered, whereupon a new association usually formed. During the intervals between periods of association, *Saimiri* rarely engaged in long feeding bouts in fruit trees, and almost never in large emergent fig trees. (There are

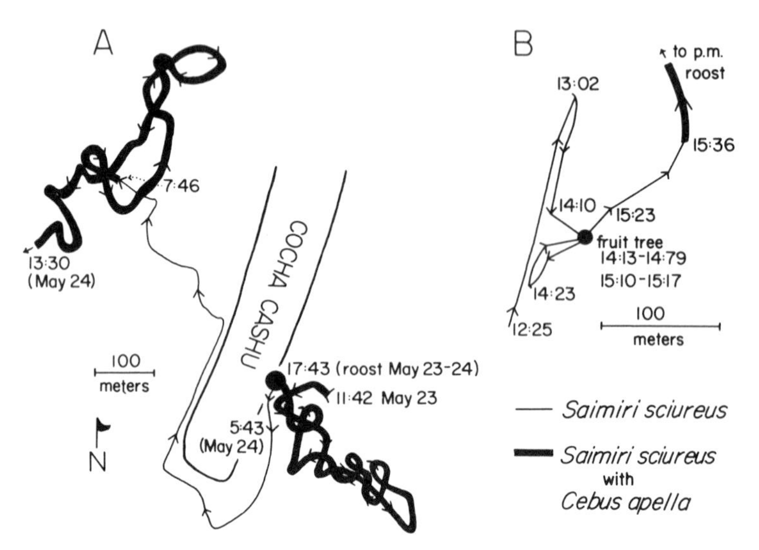

Figure 8.4 Establishment of mixed associations by *Saimiri*. A: The *Saimiri* troop under observation broke contact with one *Cebus apella* troop and quickly moved to join another. B: A *Saimiri* troop lingered along the lake front for several hours until vocalizations of *Cebus apella* were heard at about 15:20. The *Saimiri* then moved rapidly to establish contact.

seasonal deviations from this pattern that will be discussed later.) At the risk of judging the motivation behind such behavior, it seems that the *Saimiri* were "searching" for a new *Cebus* group. This impression is strongly supported by a number of occasions on which a lone *Saimiri* group was observed to respond immediately upon hearing nearby *Cebus* vocalizations. Rapid travel directly to the *Cebus* troop led to the formation of a new association.

While *Saimiri* groups tended to move to entirely new areas upon dissolution of associations, *Cebus* tended to remain in the same areas they had been using previously. In most cases the pattern of movement following a break is indistinguishable from that before. Fruit tree use also does not appear to change.

Comparison of travel rates in and out of associations show, however, that the species exert an appreciable effect on one another (Table 8.2). Unaccompanied *Saimiri* traveled nearly twice as far per hour as unaccompanied *Cebus apella*, though this is perhaps not a fair comparison. Much of the distance traversed by *Saimiri* while alone was in moving from one *Cebus* home range to another. Feeding was largely suspended. Thus the behavior cannot be regarded as truly normal or representative of the species. The comparison of *Cebus* movement rates while accompanied and alone is more likely to be free

Table 8.2
Travel rates of *Saimiri* and *Cebus apella* together and alone (m/hr.).

Time Period	Saimiri *alone*	*Both species together*	C. apella *alone*
All samples	254 (39)		
October		186 (194)	132 (42)
January		157 (61)	102 (83)
June		188 (44)	136 (33)

NOTE: Number of hours in parentheses. The two travel significantly faster together than *C. apella* does alone ($p<0.05$).

of bias because the troops continued to circulate within their circumscribed home ranges regardless of the presence of *Saimiri*. The three samples taken at widely separated times of year are consistent in indicating that *Cebus* troops travel about 40% farther per unit time when accompanied. It would be hard to construe this additional travel requirement as a benefit. Rather, it seems more plausibly to represent an imposition, and as such, a manifestation of competition or interference between the associates. The alternative explanation is that *Cebus* is speeding up its pace for the sake of remaining with the *Saimiri*, but in view of evidence already presented, and more to be considered, this appears unlikely. The interpretation that is consistent with the notion of *Cebus* as leader rather than follower is that *Cebus* is being obliged to forage farther to offset the effect of the larger food demand of a swollen group. Such effects have been documented in other primate studies. Gautier and Gautier-Hion (1969) noted that mixed *Cercopithecus* troops progress more rapidly than single troops, and Waser (1977) found a strong positive correlation between group size and daily travel distance in mangabeys (*Cercocebus albigena*).

FORMATION AND BREAKUP OF *Saimiri-Cebus* ASSOCIATIONS

The behavior of *Saimiri* when alone and just prior to the formation of associations has already been described. Active choice by *Saimiri* is, in addition, implied by the observation that associations frequently broke up during periods when the *Cebus* were stationary. On some of these occasions the observer noted an unusual amount of loud calling among the *Saimiri* just before the break occurred. The focus of leadership within a *Saimiri* troop in these circumstances is not easily discerned.

Virtually all cases in which the troop under observation switched its affiliation from one group to another involved moves of *Saimiri* between two *Cebus* troops. If *Cebus* rather than *Saimiri* were the follower species, one might expect breaks to occur near the home range limits of *Cebus* troops, but no such tendency was apparent. Moreover, *Cebus apella* troops were not observed to join preexisting associations already containing a conspecific group. This did occur with *Saimiri*, however, and huge throngs composed of two *Saimiri* and one *Cebus* troop sometimes remained together for a day or more.

Further evidence for a passive role of *Cebus* in the association is the lack of any pronounced vocal or behavioral response to the comings and goings of *Saimiri*. Usually they continued their current activity without any discernible perturbation.

The ensemble of available evidence thus points strongly to unequal roles for *Saimiri* and *Cebus* in the formation and maintenance of their associations. *Cebus* seem to exert a predominant influence in determining the pace, course, and distance of movements, including the sequence of fruit trees visited. *Saimiri*, in turn, can be seen to conform to the activity and ranging patterns of the particular *Cebus* group they have joined. Therefore, it seems probable that the benefits derived from the relationship fall principally on *Saimiri*, while the *Cebus* either gain little or lose slightly as a consequence of the increased competition. We shall now inquire further into the possible exchange of benefits and detriments between the species by examining seasonal variations in the parameters of their associations.

Seasonal Variation in Associative Behavior

Table 8.3 gives the percent of time spent in and out of association during all samples on the component species. There is clearly much variability, and the sample sizes for any given species and season are obviously small, so the patterns we think we perceive must be regarded as somewhat tentative.

Prior to discussing the observations, a comment is in order on the relative abundances of *Saimiri* and *Cebus* troops in the study area. Depending on whether or not one counts troops whose ranges only partially overlap the study area, there are from 7 to 10 troops of *C. apella* and 3 or 4 of *C. albifrons*, a total of 10–14. The number of *Saimiri* troops is uncertain because there is almost total overlap of their ranges, and because most of the troops were not individually recognizable. There were occasions on which we counted up to 6 troops in the study area at once, but at other times no more than 2 or 3 could be found. Of course there is the likelihood that some troops were overlooked in every census. A reasonable guess at the mean number would be 4 or 5. In any case, the number of *Cebus* troops is clearly greater than that of *Saimiri* troops. Thus, we cannot expect to see a given *Cebus* troop accompanied as often as a given *Saimiri* troop, even if every *Saimiri* troop in the area were associated virtually all the time. This expectation is clearly upheld by the data in Table 8.3, which show that *Saimiri* are with *Cebus* to a substantially greater degree than *Cebus* are with *Saimiri*.

Associations of the longest duration (up to 10 days or more) occurred in the late dry and early wet seasons in the period of August through November. This is a time of high and increasing fruit production by the forest, and perhaps more importantly, a time when many of the available fruits are of the soft, fleshy type favored by *Saimiri*. Later, in the middle of the rainy season (January–February), associations were markedly intermittent and of short duration. Although this corresponded to the annual peak of fruit production (Chapter 2), the *Cebus* were concentrating heavily on *Inga* pods and cacao (*Theobroma*) fruits, neither of which are especially sought by *Saimiri*.

Table 8.3
Saimiri-Cebus associations: Percentage of time per sample

		Type of Association				
As observed while following	*Sample*	*Species alone*	*Few* Saimiri *with* Cebus	Saimiri *with* C. apella	Saimiri *with* C. albifrons	Saimiri *with* C. apella *and* C. albifrons
Saimiri	Sept.	8		90	2	0
	Feb.	38		21	41	0
	May	2		85	6	7
	Aug.	35		65	0	0
	Mean	**21**		**65**	**12**	**2**
C. apella	Aug.	5	6	89		
	Oct.	17	0	83		
	Jan.	61	13	26		
	April	39	25	36		
	June	23	42	35		
	Mean	**29**	**17**	**54**		
C. albifrons	Nov.	14			86	
	March	98	1		1	
	June	55	10		35	
	Mean	**56**	**4**		**41**	

By March, *Saimiri* had virtually disappeared from the high ground forest (note the lack of associations in the March *albifrons* sample, Table 8.3). Weeks went by without a single *Saimiri* troop appearing at the house clearing, an almost daily event at other seasons. The reason for this was that large numbers of *Saimiri* had concentrated in the flooded lake bed and in the successional vegetation along the river. This habitat offered dual feeding attractions. First, there was an abundance of fruit, *Cecropia leucophaia, Xylopia* sp., and *Casearia decandra*, species that are avidly eaten by *Saimiri*. But second, and more remarkable, was the availability of a bonanza of arthropod prey. The river by then had reached its annual crest and large areas of low-lying terrain along its banks were inundated. The flooding obliged much of the leaf-litter arthropod fauna to take refuge just above the water level on cane stalks and the stems of small understory plants. Here they were conspicuous and easily harvested by *Saimiri*. For nearly two months (March–April), for as long as the fruit and flooding lasted, it was our impression that nearly all the *Saimiri* in the area were crowded into the flood zone. Many other monkeys (*Cebus, Ateles, Alouatta, Saguinus*) were also attracted by the fruit and insects, so the *Saimiri* were not always alone during this period. Nevertheless, their mass abandonment of the high ground forest indicates that access to good feeding opportunities takes precedence over their attraction to *Cebus*.

Subsequent to the depletion of fruit crops and the subsidence of the water level in the riparian vegetation, troops of *Saimiri* again began to circulate in the forest. By then (May) fruit production was at its annual nadir, and the only significant sources were widely scattered fig trees. A new pattern then developed. *Saimiri* fed heavily on figs (78% of all fruit consumed in May). Troops joined the resident *Cebus* group and remained in consort so long as the supply of figs held out. It was during this period that we sometimes observed two *Saimiri* troops with a single group of *Cebus*. As soon as the local fig had been exhausted, the *Saimiri* moved on to look for another *Cebus* troop and hopefully with it, another fig. If none was fruiting at that time within the new home range, the *Saimiri* moved again after a few days (Fig. 8.5). This behavior is also evident in the June *albifrons* sample. For the first part of the period, the *albifrons* were feeding daily at a huge *Ficus perforata*. After the figs gave out, the *albifrons* switched to eating *Astrocaryum* nuts and the *Saimiri* departed.

Given the fact that *Cebus* troops outnumber *Saimiri* troops in the study area by about two to one, any *Saimiri* troop is free to exercise a certain amount of choice in discriminating between *Cebus* troops. If improved food finding were a motivation for the joining behavior, it would profit *Saimiri* to remain in any home range only so long as the local food supply were good. When *Cebus* are using food species that are not palatable or accessible to *Saimiri*, the latter should, and apparently do, move on to other areas. And in keeping

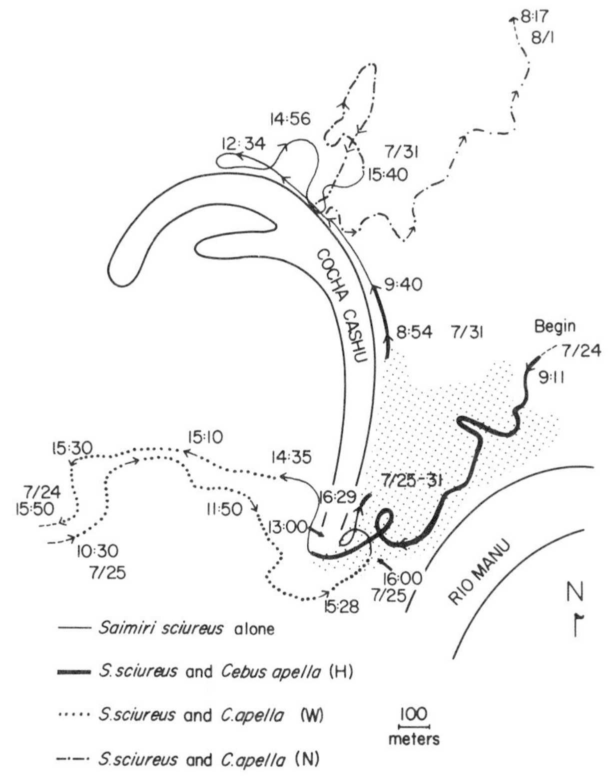

Figure 8.5 Movements of a *Saimiri* troop between July 24 and August 1, 1977. During the 8-day period, four successive associations were formed, respectively, with *Cebus apella* troops designated H, W, H, and N. Stippling indicates area covered from July 25 to 31 during the second period of association with the H *apella* troop. The straight-line overland distance between extreme points of contact was 3 km.

with this, seasonal changes in habitat use seem conditioned by shifting patterns of food availability rather than by parallel behavior on the part of *Cebus*.

Mixed Associations Including Both *C. apella* and *C. albifrons*

The data in Table 8.3 indicate that the two *Cebus* species join each other rather rarely. This happened most regularly during the period of minimum fruit abundance (May–June). Indeed, at other times of the year we did not see mixed *Cebus* troops (with or without *Saimiri*), though troops of the two species frequently passed by one another or interacted in fruit trees. During the May and June samples, however, unambiguous mixed associations occurred several times. These lasted only a few hours, except in one instance when the two troops ranged together for nearly two days. During periods of association, *albifrons* almost always lagged behind *apella* and appeared to play no role in directing the course of movement.

The circumstances of the observations suggest that *albifrons* uses *apella* to lead the way to fruit trees. Although the evidence is scanty, this interpretation is credible in light of the dominance *albifrons* exerts over *apella*. Many instances of this were seen, both when individual *albifrons* supplanted individual *apella* from feeding sites, and when the intervention of an *albifrons* alpha male resulted in the withdrawal of an entire *apella* troop. On two occasions when *albifrons* followed *apella* to food resources the behavior was particularly blatant. On the first of these, the *albifrons* troop followed a group of *apella* for about an hour, stopping to check each fruit tree visited by the *apella*. On the other occasion, the *albifrons* foraged semi-independently of the *apella* while crisscrossing their path and making repeated contacts over a period of more than six hours. At least three times during the interval the *albifrons* supplanted feeding *apella* and appropriated the trees for their own use. Here the benefit derived by *albifrons* from the interaction is clear.

Evaluation of Exchanged Costs and Benefits

Near the beginning of this chapter it was stated that unequal contributions of the species toward the formation and maintenance of mixed groups could be taken to indicate an imbalance in the exchange of benefits. In examining the evidence, we have seen that there is indeed a strong asymmetry of behavior which suggests that *Saimiri* is the active joiner while *Cebus* is ostensibly passive in its participation. One could be justified in concluding from this that *Saimiri* is the principal beneficiary of the interaction, but at this level our understanding is still quite incomplete. It is not clear what benefits *Saimiri* may derive, nor whether they are acquired at the cost of certain coincident disadvantages. Moreover, the net effect of the association on *Cebus* remains thoroughly enigmatic. To clarify these issues I shall evaluate the possible costs and benefits inherent in the interaction under the headings of predation, foraging efficiency, and competition. The evaluations must necessarily be qualitative because the trade-offs involve quantities that cannot be measured in the same units (e.g., increased predator protection vs. a disadvantage in competition). The proper comparative units are of course those of individual fitness, but obviously measurements in this fundamental currency are beyond our capability.

PREDATION

Improved predator avoidance is an often-cited reason for the existence of social groups of animals (Crook 1972; Curio 1976; Gartlan and Struhsaker 1972; Hamilton 1971; Jarmon 1974; Lazarus 1972; Morse 1977; Pulliam 1973; Vine 1971). Several kinds of observations suggest indirectly that *Saimiri* are drawn into mixed species groups in order to reduce their vulnerability to predation. One is the fact that floating subgroups of subadult or low-ranking males are not found except in the company of other monkeys. Usually they annex themselves to a *Cebus* troop, but on occasion they can be seen accompanying *Callicebus* or *Saguinus*. But beyond the possibility of mere safety in numbers, there are additional circumstances which could help explain the high degree of preference *Saimiri* shows for *Cebus* in forming mixed groups. *Cebus* have a well-developed predator alarm system. The alpha males are exceptionally alert and seem to spend more time scanning than do other troop members. Moreover, the alarm bark of *Cebus* is recognized as such and instantly responded to by *Saimiri*, as well as by other species of monkeys and even birds. (Monkeys, in turn, often jump or tense at the alarm calls of birds, e.g., caciques or macaws). Indeed, the response of *Saimiri* to *Cebus* alarms is generally much stronger than their response to their own calls. *Saimiri* often give alarm "peeps," occasionally even in chorus, without provoking much reaction from the *Cebus*. One sees this especially when mixed groups are feeding in large canopy fruit trees. The *Saimiri* may run or fling themselves out of the tree in panic, while the *Cebus* merely look up or continue their feeding.

This asymmetry of behavior might be explained in part by the size difference between the species. A predator large enough to take a *Cebus* could also take a *Saimiri*, though the converse would not always be true. Another consideration is that the frequency of "false alarms" may be higher in *Saimiri* groups, making them less reliable signalers of danger than *Cebus*, in which alarms are initiated or reinforced by the alpha male. Inherent levels of watchfulness may also differ between the species. *Saimiri* forage more intensely than do *Cebus* (Chapter 7), and as a result have less time available for vigilance. Moreover, the large groups and dominance-stratified mating system of *Saimiri* may mean that the average male has relatively little genetic stake in protecting other group members (Sherman 1977). In contrast, a dominant male *Cebus* probably has a large genetic representation in the young of its group and so may play a disproportionate role in watching for predators, as do the dominant males of some other primate species (Hall 1965).

The possible ways in which *Saimiri* and *Cebus* could interact with respect to predator avoidance are many and complex. It is thus difficult to arrive at an unambiguous judgment as to where the balance lies. From the circumstantial evidence cited above, it seems clear that *Saimiri* benefits both from the more definitive alarm calls and from the greater watchfulness of *Cebus*. How the association affects *Cebus* is not so clear. The presence of so many additional animals imposes an energetic burden in the greater distances accompanied *Cebus* troops are obliged to travel each day. There may be a compensatory advantage, however, in that the *Cebus* are likely to be the beneficiaries of a selfish herd effect (Hamilton 1971: greater protection enjoyed by individuals that are near the center of a group). Whether this over the long run would be sufficient to offset the energetic imposition of the *Saimiri* is a question that would be very difficult to answer.

One more effect of sociality that we must consider is that of expanded group size. Here, any benefits or detriments would be shared equally by the two species. There will be safety in numbers when the rate of successful predator attacks fails to increase in proportion to group size. While this has been observed to be the case in schooling fish (Cushing and Harden-Jones 1968; Neill and Cullen 1974), it can only be conjectured for primates. Even though a mixed *Saimiri-Cebus* troop will certainly have a greater detection radius than a single group from the point of view of a predator, there will be only one group available to be detected instead of two. The advantage would seem to lie on the side of coalescence, especially in the likelihood that the larger group will possess a greater aggregate vigilance (Kiltie 1981).

FEEDING EFFICIENCY: FRUIT

In the absence of any foreknowledge of their locations, fruit trees probably cannot be detected by arboreal monkeys from more than a few crown diameters away. Thus, when suitable trees are scarce, a great deal of time and energetic expense must be invested in finding them. A species that occupies a large home range, which it can cover only infrequently, might therefore gain from following another species that uses a smaller area more intensively, especially if the dietary overlap is high. This was first suggested as a possible motivation for mixed troops by Gartlan and Struhsaker (1972). The gross behavior of *Saimiri* and *Cebus* is certainly consistent with this interpretation, but there is at least one detail that does not seem to fit. It is the tendency of *Saimiri* troops to remain associated with a single *Cebus* troop for long periods during the early rainy season peak of fruit abundance. If the *Saimiri* merely wanted to learn where the best fruit trees were, they could do so in the first few days of association. After that, one would suppose that their interest would lie in visiting the most preferred trees on their own schedule, ideally ahead of their larger compatriots (as African red-tailed monkeys seem to do: Waser and Case 1981). Instead, they prefer to remain continuously with the *Cebus*, even though by doing so they forfeit their choice of routes and feeding stations and are subject to being expelled from many of the best trees. All this implies that the interest of *Saimiri* in *Cebus* extends beyond the mere discovery of ripe fruit crops.

There is, however, an unambiguously direct foraging benefit that *Saimiri* gains from *Cebus*. It is the opportunity to exploit the remaining mesocarp of *Scheelea* nuts that have been opened by *Cebus* and partially eaten. These nuts come in large tightly packed clusters that require as much as half an hour of work by the dominant male *Cebus apella* to open. Once the cluster has been breached, the individual nuts themselves present a second challenge because the rough outer husk must be stripped away to expose the edible coconut-like mesocarp. None but the largest male *Saimiri* are capable of doing this. However, the question of ability is moot because the *Cebus* totally dominate the cluster, often until it is entirely finished. While avid in their use of these nuts, apparently because of their high caloric and oil content (Hladik et al. 1971), the *Cebus* are not meticulous feeders and allow many nuts to fall to the ground before even half of the edible material has been consumed. Waiting *Saimiri* then dash down to snatch up the unfinished nuts. While the transfer of a benefit is obvious, the rate at which *Cebus* drop the nuts is sufficient to occupy only a small part of a *Saimiri* troop. Moreover, the discovery of ripe *Scheelea* clusters is an occasional event at best, too infrequent to account for the year-round interest of *Saimiri* in *Cebus*.

INSECT FORAGING

Since both *Saimiri* and *Cebus* spend a major part of their lives searching for prey, it may be that the association in some fashion facilitates the foraging of one or both species. The disturbance created by so many animals moving through the foliage, for example, could flush hidden prey that could then be harvested, a suggestion offered by Gartlan and Struhsaker (1972) and by Rudran (1978). That this should work for monkeys seems very unlikely in view of what we learned in the last chapter. Large mobile prey are best captured by stealth or surprise; once startled they can easily bound or fly away from an oncoming primate. Instead, it is birds with the ability to capture prey on the wing that profit from this situation. Five species routinely associate with primates at Cocha Cashu for just this purpose: *Harpagus bidentatus* (Accipitridae), *Crotophaga major, Neomorphus geoffroyi* (Cuculidae), *Monassa nigrifrons* (Bucconidae), and *Dendrocincla fuliginosa* (Dendrocolaptidae); several other species attend troops on a more casual basis. The larger the troop, the more birds are likely to be present, just as large army ant swarms attract more birds than small ones (Willis and Oniki 1978). The gathering of hawking birds at major disturbances (e.g., prairie fires, storm fronts) is a well-documented phenomenon, and the opportunistic use of primates as beaters for certain birds has been noted in Neotropical (Greenlaw 1967; Fontaine 1980), African (Chapin 1939), and Asian (Stott 1947; Oppenheimer 1977) forests. It thus seems that the principal beneficiaries of the effects of disturbances are birds, and that primates neither gain nor lose in their contributory roles.

A more direct form of interaction during foraging has been noted by Klein (pers. com.), who observed *Cebus apella* pirate large prey items from *Saimiri* through aggressive intimidation. We have not recorded any instances of such behavior in well over 1,000 hours of observation of mixed troops at Cocha Cashu. It thus may be so rare as to be of little consequence. In any case, the frequent intimidation of *Saimiri* in fruit trees fails to deter their persistent following of *Cebus*.

COMPETITION

A major effect of bringing together individuals or groups of animals is to make competition more direct, not to reduce the total demand of the organisms on their resource base. While direct competition (i.e., via aggression) may entail a greater real cost to an individual than indirect competition (food depletion in the absence of interference), it cannot be assumed that this is always true. For example, Cody (1971) suggested that by forming mixed flocks, birds could improve their mutual foraging efficiency by together depleting the resources in certain areas and then avoiding those areas subsequently.

Both direct and indirect competition are involved in the *Saimiri-Cebus* interaction. Direct competition is frequent in fruit trees as *Cebus* displace *Saimiri* from their chosen feeding stations. Although the net effect of this on the feeding success of *Saimiri* is difficult to assess, individual monkeys are often excluded from trees for periods of several minutes. Indirect competition is more difficult to demonstrate. The fact that *Cebus* often do exert the effort to expel *Saimiri* does suggest that it occurs in some fruit trees. The more tolerant attitude that is displayed on other occasions may just reflect an excessive cost in chasing the more numerous and agile *Saimiri* completely away from a feeding area. A second indirect effect of *Saimiri* on *Cebus* was seen also in the lengthened daily travel distance of mixed troops.

It is conceivable that direct competition could benefit both *Saimiri* and *Cebus* if their joint movements were coordinated in such a way as to regulate the intervals at which the troop returned to particular foraging areas in the manner proposed by Cody (1971). However, we have no evidence that the monkeys actually do this. Returns to individual hectares within the home range are irregularly spaced in time, and certain areas, such as the lake margin, are searched with a conspicuously greater frequency than others. The situation is undoubtedly complicated even further by the fact that the sites chosen for insect foraging are not independent of the locations of the fruit trees currently being exploited. As Cody rightly emphasized, regulation of return times entails an advantage when the resource being exploited is self-renewing over a definite recovery period. Therefore his model could not be expected to apply to synchronously ripening fruit crops because these are effectively nonrenewing resources. Evidence for regulation of return times is thus nonexistent. Instead, our observations indicate that the interaction leads to competitive consequences that are negative for both species. *Cebus* is obliged to travel farther and to contest *Saimiri* over the possession of fruit trees, while *Saimiri* experiences an appreciable amount of interference in its attempts to gain access to good feeding sites. We conclude that mixed associations are favored, at least by *Saimiri*, in spite of the disadvantageous effects of direct and indirect competition.

Saimiri-Cebus Associations: Conclusions

We have seen in the last section that the relationship between *Saimiri* and *Cebus* entails undeniable, though not easily measured, costs to both species. These costs take the form of lost or deferred feeding time, and, for *Cebus*, an increased requirement for travel. Nevertheless, a great deal of evidence indicates that the mixed associations are initiated and maintained through active behavior on the part of *Saimiri*.

What then are the benefits that *Saimiri* receives from the association? We have identified three: (1) occasional access to half-eaten *Scheelea* nuts; (2) improved discovery of resources through parasitizing the more current and detailed knowledge *Cebus* has of the locations of ripe fruit crops within its home range; and (3) better warning against predators provided by the greater vigilance and more authoritative alarm bark of *Cebus*. Access to *Scheelea* nuts can be discounted as an important contributory factor because of the infrequency with which ripe nut clusters are discovered, and because of the fact that only a few members of a *Saimiri* troop are able to scavenge nuts on any occasion. Enhanced discovery of fruit resources is likely to be a much more powerful motive for association. But, as pointed out above, most of the gain through this mechanism would be realized in the first few days. Later, it would seem that *Saimiri* could profit better from their newly parasitized knowledge by traveling on their own and avoiding interference from *Cebus*. Moreover, this explanation fails to account for why the associations are most durable during the season of maximum fruit abundance. It does, however, account for the frequent switching from one *Cebus* troop to another during the period of fruit scarcity.

Of the three proposed benefits, the only one that is consistent with the year-round interest that *Saimiri* shows in joining *Cebus* is the opportunity to eavesdrop on *Cebus* alarm calls. This I believe to be the principal motive. The joining behavior is then modified during periods when resources are patchy or in short supply by more frequent switches. An additional benefit is then gained when *Saimiri* is led to fruit trees. To show directly that *Saimiri* gains more in improved predator protection and more efficient discovery of resources than it loses in its competitive subordination to *Cebus* would require a lifetime in the field. Thus, while the conclusion is inferential, it is probably as strong as it could be under the circumstances.

While little doubt can remain that *Saimiri* receives a benefit from the interaction, what about *Cebus*? Its passive role in the maintenance of associations suggests that the net effect of the presence of *Saimiri* is neutral or negative. I suspect that it is slightly negative. The expanded group size imposes a greater daily travel requirement, and the presence of persistent throngs of *Saimiri* at many fruit stops is a nuisance at the very least. On the positive side it is hard to make much of a case. The operation of a ''selfish herd'' effect seems the best possibility. It would be credible if *Saimiri* and *Cebus* shared the same predators. We have seen *Saimiri* but not *Cebus*, attacked on a number of occasions by *Spizaetus ornatus* (ornate hawk eagle). Harpy eagles (*Harpia harpyja*) and Guianan crested eagles (*Morphnus guianensis*) have been seen to attack both *Cebus* and *Saimiri* (cf. Chapter 9). *Spizaetus* (ca. 1000 g) is probably too small to be much of a threat to *Cebus*, while most of the recorded prey of *Harpia* (4–6 kg) are larger than *Saimiri* (Rettig 1978). *Morphnus* (1750 g) is intermediate in size and thus may threaten both equally. The existence of shared predators does suggest that *Cebus* benefits by being surrounded by a crowd of *Saimiri*; but in the absence of any further evidence, it seems in balance that *Cebus* is being mildly parasitized by *Saimiri*.

Finally, a word about the interaction of the two *Cebus* species. True associations, as distinguished from brief encounters, were observed only during the season of minimum fruit abundance. *Albifrons* then sometimes followed *apella* for periods of up to two days, using many of the same fruit trees and supplanting *apella* from a few of them. These observations, combined with the fact that the home range of *albifrons* is much larger than that of *apella*, imply that *albifrons* uses *apella* to lead it to resources which it can then control through aggressive dominance. If this is indeed so, why then are the associations so seasonal and infrequent? One possible reason is that the indirect (passive) competition experienced in so large a group of monkeys must be very great. This would be especially true in the presence of a horde of *Saimiri*. Under most circumstances, better feeding could be enjoyed by being alone. But there is perhaps another reason. Although an adult male *albifrons* can easily intimidate all but his counterpart in an *apella* troop, there is still a possibility that a male *apella* could retaliate against some less prepotent member of the *albifrons* troop. It may be that the latent threat of serious aggression coupled with the disadvantages of a doubled food demand keeps the two species apart except when resources are so scarce that *albifrons* cannot otherwise locate them.

Mixed *Saguinus* Associations

The relationship of the two *Saguinus* species differs from that of *Saimiri* and *Cebus* in a number of respects. The groups are strongly territorial and the associations are permanent, rather than temporary or opportunistic. The number of groups of each species in the study area is approximately the same, allowing a nearly perfect one-to-one matching of partners. Instead of having different ranging habits, associated groups share a common territory within which their ranging is closely parallel. Being very nearly equal in size, the two species are probably subject to the same set of predators, and unlike *Saimiri* and *Cebus*, they respond in apparently reciprocal fashion to each other's alarm calls. Vocal communication between the species is frequent and serves to coordinate their movements, especially when associated groups are out of sight of one another. And in perhaps the largest departure from the *Saimiri-Cebus* model, interest in joining and following appears to be mutual, with both species participating to about the same extent.

The two systems are alike in that the associated species are close ecological relatives having similar diets, activity budgets, and movement rates. In providing a necessary compatibility of life styles, such parallels are undoubtedly a precondition for any strong mixed foraging interaction. Another similarity is the unambiguous dominance of one species—here, *imperator* over *fuscicollis*.

Our efforts to understand the costs and benefits inherent in the *Saguinus* interaction have not progressed as far as they have in the case of *Saimiri* and *Cebus*, and further work is underway at this writing. Nevertheless, I shall attempt to outline the major possibilities and to evaluate them from the perspective of current knowledge. Even though we shall not be able to come to a definite resolution of the issues, it will be clear that the incentives underlying the *Saguinus* relationship are not identical to those that promote the joining of *Saimiri* and *Cebus*. This shall be apparent at the outset as we examine the mechanics of maintaining the mixed association.

Maintenance of the Alliance

Because associations of *Saguinus* groups are long-lived, one does not have the opportunity to see how they are formed, only how reunions are affected after brief periods of separation. Associated groups are often out of sight of one another; indeed, it is our impression that this may be the case more than half the time, though we were not able to quantify the point. Activities are frequently nonsynchronous. Insect foraging is usually conducted separately, as are rest sessions. Even during synchronous rest periods, the two groups are seldom in the same crown. They may be in nearby crowns or as much as 100 m or more apart. Sometimes one family will rest low beneath a protective canopy of foliage, while the other is 30 to 40 m up in a vine-shrouded treetop. Yet in spite of the considerable independence of action, associated groups follow roughly parallel courses through the day, traversing the same parts of their territory and coming into visual contact repeatedly. Activities tend to be coordinated most closely during feeding when the groups are visiting a succession of fruit trees or *Combretum* vines. Then, depending on the size of the tree or vine being exploited, the two groups may feed together, or in sequence, or in parallel if another plant is present in the vicinity. Movements between feeding stops tend to be well synchronized, with the groups moving along together or in file. To document these assertions adequately, it would be necessary to have observers follow and map the two groups simultaneously. This has not yet been done, though it is planned for the future.

During periods when associated groups are out of visual contact, periodic exchanges of vocalizations serve to inform each of the other's whereabouts. The vocalizations used are loud and species-specific, and to our ears are indistinguishable from those used to communicate with an isolated family member or with neighboring conspecific groups. Either species may initiate a vocal exchange. On most (but not all) occasions the other species will answer. Such exchanges occur at the start of every day as each group leaves its roost tree, and intermittently at other times, including rest periods when the families are concealed in separate retreats. In many instances there is no obvious response to a query other than the answer. Resumption of coordinated travel at the conclusion of a rest session is most often brought about by one group's going to the other's rest site. Either species may be the instigator. Then, when the two groups move off together, either species may take the lead, though *fuscicollis* assumed this role more frequently, perhaps only as a consequence of its more rapid cling-and-leap mode of locomotion (Table 8.4).

Just as in the case of *Saimiri-Cebus* associations, we were able to take advantage of some unusual circumstances that arose fortuitously in the course of the study to resolve the question of which species was the active joiner. I have already asserted that in the *Saguinus* relationship both species are active participants. The most convincing evidence that *imperator* intentionally joins

Table 8.4
"Leadership" of mixed *Saguinus* progressions.

Sample	*No. observations*	S. imperator *ahead*	S. fuscicollis *ahead*
April	20	3	17
July	32	12	20
Both	52	15	37

NOTE: $p << 0.01$ that leadership is random.

fuscicollis came from periods when our tame *fuscicollis* troop was occupying the territory of an intractably wild *imperator* family. Up to several times a day the *imperator* group attempted to approach after first making vocal contact with the *fuscicollis*, but as soon as the observer was detected, the *imperator* fled into the distance. On most of these occasions, the *fuscicollis* made no serious effort to follow.

That *fuscicollis* intentionally joins *imperator* was demonstrated by quite a different situation. For a period that encompassed the April and June 1977 samples, it happened that there was a vacancy in the territorial mosaic. The number of *imperator* groups in the study area was one greater than the number of *fuscicollis* groups. It then developed that our *fuscicollis* study troop began to use the vacant territory as well as the territory it had been using during the previous sample in November 1976. The one *fuscicollis* group was then allied, alternately, with two mutually hostile *imperator* groups. It followed the practice of associating with one *imperator* group for several days, and then switching to the other (Fig. 8.6). When transiting between territories the *fuscicollis* called repeatedly until they obtained a response from the resident *imperator*, whereupon the two soon joined. There thus remains little doubt, at least in our view, that *fuscicollis* and *imperator* both actively participate in coordinating their associations.

Behavior In and Out of Association

The shuttling back and forth of the *fuscicollis* troop between *imperator* territories provided us with a remarkable opportunity to study the behavior of each species alone and in association in the same time and place. Such opportunities must be very rare, because at no other time in the five years we have been studying primates at Cocha Cashu have we seen a comparable situation. In their comings and goings the *fuscicollis* graciously provided us

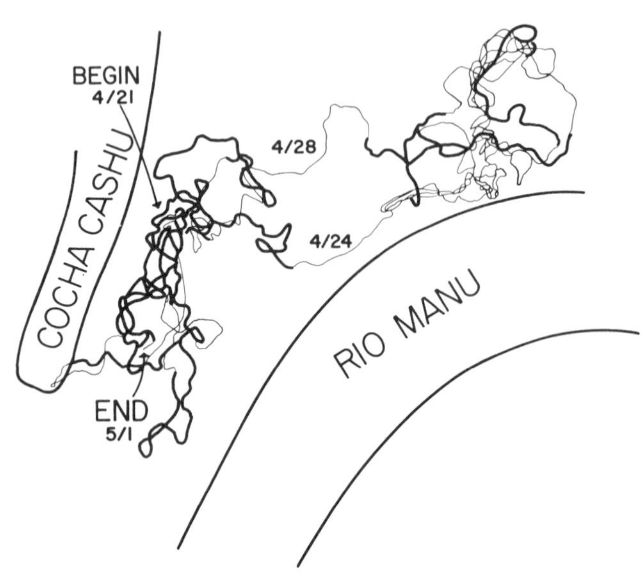

Figure 8.6 Movements of the *Saguinus fuscicollis* study troop within and between the territories of the S and SE *imperator* troops. Thin line: *fuscicollis* alone; heavy line: *fuscicollis* with *imperator*.

with very nearly equal sample periods for the together and alone conditions (Table 8.5).

Having seen earlier that the presence of *Saimiri* imposed a substantial increase on the travel budget of *Cebus apella*, we can wonder whether the *Saguinus* similarly affect one another. The data in Table 8.5 suggest that there is such an interaction, but there are reasons for being guarded in interpreting the results, especially the time spent by *fuscicollis* alone. While the *imperator* group remained continuously within its territory, the *fuscicollis* traveled back and forth between territories. During the transits they tended to move faster, just as *Saimiri* did when moving between *Cebus* troops. Moreover, most of the time recorded for *fuscicollis* "alone" was within the southeast territory where the *imperator* group was extremely shy and fled at the sight us. Although we could not always be sure of what was happening, it seems likely that the *fuscicollis* were at times attempting to join the wild *imperator* because there were periodic vocal exchanges as well as occasional fleeting encounters. Our presence then was artificially promoting the "alone" status of the *fuscicollis*. Finally, there was still another factor that contributed to the high *fuscicollis* travel rate in the July sample. This was a prolonged territorial confrontation which took place on one of the "alone" days and which was accompanied by an unusual amount of running and chasing. If we were to delete from the samples all the time spent in territorial conflict, or

Table 8.5
Distance traveled and mean number of feeding trees visited per hour by *Saguinus imperator* and *S. Fuscicollis* when alone and together.

	Distance traveled per hour(m)				*No. trees visited per hour*			
	S. imperator		S. fuscicollis		S. imperator		S. fuscicollis	
Sample	*Alone*	*with* S. fusc.	*Alone*	*with* S. imp.	*Alone*	*with* S. fusc.	*Alone*	*with* S. imp.
April	93(34)	122(22)	124(22)[a]	103(44)	1.3(51)	1.6(30)	1.2(31)	1.3(43)
July	84(34)	108(22)	146(28)[a]	103(36)	1.2(33)	1.0(21)	1.6(28)	1.0(36)
Both	89(68)	115(44)	136(50)[a]	103(80)	1.3(84)	1.4(51)	1.4(59)	1.2(79)

NOTE: Number of hours of observation in parentheses. Only whole days were included in the travel samples while partial days were also included for the tree visits.
[a] Values probably inflated as described in text.

in the southeast territory, or transiting between territories, there would be no sample left. The figures as presented are thus almost certainly higher than they should be.

The data for *imperator* alone and for the two species together can probably be considered representative and as free of bias as possible. The comparison is of the *imperator* study troop in its territory with and without the *fuscicollis*. There is a suggestion of a difference in that the *imperator* in the company of the *fuscicollis* traveled about 30% more than they did while alone. Curiously, the mixed group seemed to travel somewhat faster when the species being tracked was *imperator* then when it was *fuscicollis*, though the number of sample days is small and the difference is doubtfully significant.

Just why the mixed group should travel more than the *imperator* do when alone is not obvious. It would be reasonable to expect that the added food demand of the second troop should lead to an increased number of trees visited, but the results show no hint of such an effect. Apparently, the frequent practice of feeding in parallel in nearby or adjacent trees effectively removes this form of interference. The two species thus seem to be able to monitor each other's feeding activities without incurring any measurable cost in extra feeding trips. It is still possible, however, that the average distance traveled between trees visited increases in association because the second troop frequently eliminates nearest neighbor trees as potential next stops in the itinerary. There are other possibilities as well, and at present we cannot distinguish among them. Maintaining the association could in itself lead to increased travel. This might happen as a consequence of the fact that partner groups frequently wander apart, often in connection with their separately pursued insect-foraging activities. Reestablishment of contact after separations would call for additional travel. Still another possibility is that territorial battles are more active or prolonged when both species are present. The furious running about that accompanies these occasions can add substantially to daily travel distances. The data presently at hand are simply not sufficient to control for all these separate possibilities, so the issue shall remain unresolved. The lack of increased visits to fruit trees nevertheless stands as an unexpected and interesting result to which we shall later return.

Mixed *Saguinus* Associations: Possible Costs and Benefits

Our consideration of *Saimiri-Cebus* associations in the first part of this chapter led to the conclusion that there are strong asymmetries in the relationship. *Saimiri* seems to benefit in at least three ways, while *Cebus* questionably receives any benefit at all, and appears to be put to a slight disadvantage in its foraging. From what we have been able to discern about the interaction of the two *Saguinus* species, there appears to be no equivalent asymmetry in their relationship. Indeed, the indications point to an alliance of mutual interest, notwithstanding the aggressive dominance of *imperator* over *fuscicollis*. Although the very existence of the interaction implies an exchange of benefits (or a mix of benefits and detriments), the nature of what is exchanged remains quite obscure. Nevertheless, as we shall see in the ensuing discussion, the available evidence does seem to rule out two and perhaps all three of the benefits that *Saimiri* receives from *Cebus*.

Predation. If the two species help each other watch for predators, there would be an approximately equal exchange, because each is sensitive to the other's alarm calls. However, there are other circumstances that remain unexplained if this mechanism were to be accepted as the compelling force behind the interaction. One is the fact that the associated groups spend much of their time apart, or at least out of visual contact. Whenever their separation is more than 10 or 20 m (approximately the limit of visibility in the 4–10 m zone in which they travel), the potential for reciprocal warning of predator approach is reduced. Many of the predator alarms or unsuccessful predator attacks we have observed occurred while a group was feeding in a tall exposed crown. If individuals were seeking safety in numbers, or taking advantage of the extra eyes and ears provided by additional group members, one would expect a high degree of mutual tolerance during feeding sessions. Yet, as we have already described, this is not the case. Feeding is more often sequential or parallel than simultaneous. Thus the two groups do not behave in such a way as to maximize their ability to reciprocate in predator warnings. Nor would an explanation of the association based on mutual relief from predation account for the sharing of territories. Overlapping ranges could serve this function just as well, as the *Saimiri-Cebus* situation illustrates.

Feeding efficiency. While *Saimiri* stands to gain by following *Cebus* in being led to unknown sources of fruit and in having the occasional opportunity to scavenge *Scheelea* nuts, neither of these motives could apply in the *Saguinus* interaction. The species have identical territories within which the locations of resources are presumably well known to both. Moreover, being nearly equal in size, neither possesses physical capabilities that give it access to plant resources that are not available to the other. This statement, of course, does not apply to insect foraging in which different physical capabilities are manifest. But this in itself assures that the possibilities for social facilitation in the discovery of prey are nil. Evidence for enhanced foraging or feeding success as a factor in the relationship is thus entirely lacking.

Competition. While competitive interactions in *Saimiri-Cebus* associations are negative from the points of view of both species, it is not so obvious that this is the case in *Saguinus*. The situation is clearly more complex, and the critical observations needed to unravel all the details have not yet been made. Nevertheless, we may still gain some understanding by exploring the possibilities, even if some of the arguments are largely speculative.

Direct competition between the species definitely occurs in the form of aggressive exclusion of *fuscicollis* from feeding trees by *imperator*. In spite of this, we have seen that *fuscicollis* frequently joins *imperator* under circumstances that suggest conscious, premeditated volition. It seems proper to conclude then that, whatever the cost of *imperator* aggression is to *fuscicollis*, it is offset by some larger benefit. Likewise, the cost (or threat) of indirect *fuscicollis* competition does not deter *imperator* from actively joining its associate.

Even though the "cost" of direct competition is borne unequally by *fuscicollis*, the possibility still exists that the competitive effects that each exerts on the other largely cancel out. This could come about through the tendency of *fuscicollis* to run ahead and arrive first at feeding trees. By feeding quickly, the *fuscicollis* could visit the best spots before the *imperator* arrived and expelled them. Or, alternatively, depending on which group arrived last, one or the other might elect to go to another slightly less desirable feeding tree nearby. Thus, though the *imperator* always have the option of expelling the *fuscicollis*, the latter may more often reach the best feeding sites first, thereby canceling the disadvantage of being subordinate.

A possibly important implication of feeding in parallel is that it may partly nullify the indirect competition that results from increased group size. Our best evidence for this was the lack of any noticeable increase in the rate of tree visitation when the two species were in association. An increase in travel, however, does imply that the interaction entails a cost, albeit not a large one. The crux of the problem lies in the contrast between what we actually observe and something we cannot observe—how the species would behave if there were no association and if they operated entirely independently of one another in overlapping territories. By traveling together each species is potentially able to monitor the feeding activities of the other, and thereby to avoid futile visits to trees that had just been exploited. The advantage to be realized in avoiding redundant visits depends critically on the characteristics of the resource in question. They would be minimal for large, synchronously ripening (nonrenewing) fruit crops, and maximal for small, slowly ripening (self-renewing) crops that are relatively little exploited by other (third-party) species. We saw in Chapter 6 that the major resources of *Saguinus* conform to the latter set of characteristics.

Here indeed is a situation in which a mechanism something like Cody's (1971) might operate. By traveling together the two species could reduce their mutual competitive interaction and enhance their realized feeding efficiency by choosing separate but nearby trees. A further potential benefit could be gained by regulating return times to particular trees or portions of the territory. The observed habit of traveling in separate but parallel paths is entirely consistent with this mechanism, but as mentioned above, it does not maximize the potential for reciprocal warnings against predators. Predator warnings could nevertheless occur as a consequence of the proximity of the groups, providing an additional reinforcing benefit. There would also be a more indirect way by which exposure to predation might be reduced. By cutting down on unnecessary travel and unrewarded visits to fruit trees, the practice of cooperative competition would lead to an increased amount of time available for hiding (rest) in predator-proof retreats, as well as to energetic savings.

These arguments would apply to the major portion of the year in which the supply of fruit and nectar resources in the territory is adequate for both troops. But might competition between the species not become intense if plant resources were in short supply? Under such extreme circumstances one species might be expected to eliminate the other. There are a number of ways to weasel on this issue, and any or none of them might be correct. It could be that in the absence of sufficient plant resources the tamarins could increase their consumption of arthropods as *Saimiri* do, though we have no evidence for this. In such a case, their different foraging techniques would alleviate the strains of competition. Alternatively, it is possible that some territories are inherently better for *imperator* and others for *fuscicollis* so that the outcome of knock-down resource competition would be a mosaic of survivorship that favored neither species decisively. Or, it could be that the system of joint territoriality operates in such a way as to insure both partners of an adequate resource supply in all but the most extraordinary years. In our present state of ignorance we are simply not in a position to evaluate any of these possibilities.

A final comment is due on the extraordinary phenomenon of joint territoriality. Its net effect on the participants is an issue that lies beyond our empirical reach because there is no opportunity to compare the same species as they might behave under a system of overlapping instead of coincident territories. A possibility we have not yet considered is that each group may aid the associated group in the maintenance, and perhaps even in the expansion, of the joint territory. Successful groups could conceivably increase their holdings to the point that a portion of the territory could be "budded off" to permit the entry of offspring into the reproductive population. Such a budding process has been documented in the Florida scrub jay (Woolfenden and Fitzpatrick 1978). Small or unassociated groups might be forced to give ground and ultimately to abandon their territories altogether. These are open possibilities that must remain unresolved at present. Observations of territorial confrontations involving two groups of one species (*imperator*) and one of the other (*fuscicollis*) are inconclusive. The odd group seemed to take no sides in the proceedings; but then there would have been no advantage for it to have done so, because it had free access to both territories anyway (cf. Fig. 8.6). More extended observations on known individuals and groups will be required for further progress.

Mixed *Saguinus* Associations: Conclusions

There are so many gaps in our knowledge of important details of the *Saguinus* interaction that definite conclusions about the balance of costs and benefits are not yet possible. Nevertheless, it seems abundantly clear that the *Saimiri-Cebus* model provides few insights into the very different circumstances of the *Saguinus* alliance. While a desire to seek safety in numbers and to take advantage of *Cebus* alarm calls seems to motivate *Saimiri*, one cannot easily make the same claim for the tamarins because they are apart so much of the time. Instead, it is difficult to avoid the intuitive feeling that what is important to the tamarins is the integrity of their territories and the careful husbanding of the contained plant resources. If their coordinated travel were indeed to have the effect of minimizing the number of fruit trees visited by maximizing the payoff realized in each one, there could be two derived benefits: reduced energetic cost and reduced exposure to predation while feeding in high emergent crowns. For the 80 or 90% of the year when food is amply abundant in the territories, a slight improvement in the energetic efficiency of harvesting it is probably of little consequence. Thus the important benefit may be in reducing exposure to predation, because it is one that can be realized at all times. But whichever of the possible benefits is really the more crucial, both depend on a close and continuing association and consistent joint use of the same space. A system of overlapping territories would not provide these benefits to the same degree because each group would be unable to monitor simultaneously the feeding activities of the two or more groups of the other species with which it overlapped. Cooperative competition thus does provide a mechanism for the evolution of joint territoriality. Nevertheless, the question still remains as to whether there are further implicit benefits in the potential for mutual reinforcement or expansion of territorial boundaries.

Summary

In this chapter we have reviewed evidence helping to clarify the behavioral interactions involved in two kinds of mixed species associations, and have evaluated the potential costs and benefits of the associations to the participants. Although in most cases the inherent costs and benefits cannot be measured directly, they can sometimes be inferred from some of the circumstances of the associations: whether participation is active or passive, which species lead and which follow, how the frequency and duration of associations vary seasonally, and how much the behavior of each species is altered by its participation.

Saimiri-Cebus Associations

Within the study area, *Cebus* troops outnumber *Saimiri* troops by about two to one, and accordingly, *Saimiri* are observed more often to be with *Cebus* than *Cebus* are with *Saimiri*. A great deal of evidence implicates *Saimiri* as the active participant, responsible for both initiating and dissolving the associations. Unions may last anywhere from a few hours to ten days or more, in a pattern that varies seasonally. By all appearances, the attitude of *Cebus* is one of passive resignation to the presence of *Saimiri*.

We have identified three kinds of benefits that *Saimiri* derives through the interaction. One is surely trivial. It is realized on the rather infrequent occasions when a *Cebus* group (usually *Cebus apella*) discovers and feeds on a cluster of *Scheelea* nuts. These nuts are large and well protected, and are invulnerable to *Saimiri*. However, after individual nuts have been opened and partially eaten by *Cebus*, *Saimiri* eagerly scavenge whatever remains. Another benefit follows from the fact that *Cebus* home ranges are much smaller than those of *Saimiri*. Whenever a *Saimiri* troop comes into a region it has not recently visited, it can use the resident *Cebus* to lead it to the best available fruit trees. This interpretation is supported by the observation that during times of high abundance, associations are stable for relatively long periods; but when fruit is scarce, *Saimiri* frequently switch from one *Cebus* group to another, staying only when there happens to be a fruiting fig in the current *Cebus'* home range. The third benefit we judge to be the most compelling, because it is the only one that explains the persistent affinity that *Saimiri* shows for *Cebus* even during periods when resources are superabundant. It is that *Cebus* provide a better predator warning system than the *Saimiri* appear to possess themselves. *Cebus* alarm barks produce an instantaneous response in *Saimiri*, and *Cebus*, particularly the alpha male, are more alert and spend more time scanning. The more intensive foraging of *Saimiri*, much of it done with the eyes focused less than 20 cm away, leaves them especially vulnerable to surprise attack.

There is little suggestive evidence that *Saimiri* reciprocates in the interaction to confer advantages on *Cebus*. The best possibility for this is that the large numbers of *Saimiri* may at times provide a protective cover for the more centrally located *Cebus* when the mixed group is under predator attack.

Association entails definite disadvantages for both participants. One type of disadvantage results from the fact that a mixed group contains approximately twice the biomass of either group alone. This means that there is a great deal more jostling for position in fruit trees, and perhaps more redundant searching of branches and leaves during insect foraging as well. But whatever the precise cause, the presence of *Saimiri* obliges *Cebus* troops to increase their daily travel distance by about 40%, regardless of the season. The *Saimiri* also pay a price of participation in that they are frequently expelled from fruit trees and forced to wait until the *Cebus* have finished. Nevertheless, the active part played by *Saimiri* in initiating and maintaining the associations strongly implies that the benefits derived from the interaction outweigh the costs. This is not true for *Cebus*, however, in that the association provides no definite advantage to offset the energetic burden of an increased travel budget. It thus appears that *Saimiri* is a mild parasite of *Cebus*.

Mixed groups containing both *Cebus* species (with or without *Saimiri*) occurred mainly during the May and June samples when fruit production by the forest was at its annual minimum. On these occasions *albifrons* clearly followed *apella*, and several times supplanted them from fruit trees. In a more blatant and direct way than *Saimiri*, *albifrons* is thus also able to parasitize *apella*. The common denominator in the two situations is that the pirating species have much larger home ranges than the victim.

Mixed *Saguinus* Associations

The two species of *Saguinus* at Cocha Cashu live together in permanent mixed associations consisting of one family group of each species. Allied groups share and jointly defend a common territory against similar mixed groups in the adjoining territories. Within a territory the two resident groups travel separately over roughly parallel paths, coordinating their movements fairly closely when visiting a succession of fruit trees, and more loosely at other times. For much of each day they are out of visual contact while continuing to maintain an awareness of each other's whereabouts through frequent vocal exchanges. The two species appear to participate equally in initiating vocal contacts and in reestablishing close proximity after periods of separation. Though either species may lead progressions, *fuscicollis* most frequently goes ahead. This means that *fuscicollis* often arrives first at fruit trees, but when it does so it is subject to being expelled by the dominant *imperator*. As opportunity permits, the two groups commonly feed in parallel in adjacent or nearby crowns.

A fortuitous set of circumstances provided us with a rare opportunity to study the activities of both species in and out of association. The opportunity arose when a *fuscicollis* troop disappeared, leaving an *imperator* troop alone in its territory. The *fuscicollis* study troop responded to this situation by periodically moving into the vacant territory for several days at a time and then returning to its own adjacent territory. These alternations continued for a period that encompassed the April and July 1977 samples. During this time the *fuscicollis* troop effectively maintained a dual alliance with two mutually hostile *imperator* troops. This gave us the chance to follow the *imperator* study troop on days when it was alone and on days when it was accompanied by the *fuscicollis*. During association the distance traveled per hour by the *imperator* increased by about 30%. However, the number of trees visited per hour did not increase noticeably for either species. This apparent lack of an effect of increased group size can perhaps be explained by the common tendency of the two groups to feed in parallel in nearby crowns. By this means, the potential for interference between them is reduced, while at the same time each group is potentially able to monitor the other's feeding activities.

The question of what advantages the partners derive from the interaction is still largely unresolved. Reciprocal warning of predator attacks is a possibility, but the fact that associated groups spend so much time apart suggests that this is not the only compelling motivation. An additional possibility is that of cooperative competition. This would operate through a mechanism similar to the one proposed by Cody (1971) in which species using the same pool of resources gain an advantage by exploiting them together and regulating return times to maximize the yield on each visit and to minimize the number of unrewarded redundant visits. The greatest benefit is obtained when the resources are scattered, slowly renewing, and little used by other (third-party) species.

These properties are characteristic of the major resources of *Saguinus*. Benefits would accrue in the form of energetic savings through reduced numbers of visits to empty fruit trees and of reduced exposure to predators through decreased time spent in the high canopy. Were such benefits to be realized, they would provide a satisfactory basis for understanding the extraordinary system of joint territoriality shown by *Saguinus*. A system of overlapping territories, with or without association of the respective species, would not provide the same advantages because no group could simultaneously monitor the activities of two or more groups of another species in its territory.

Whatever costs may be implicit in the interaction—for example, increased travel, interference, and aggression in fruit trees, etc.—they cannot be evaluated independently of the realized benefits. To do so would require comparison with the nonexistent alternative situation of overlapping territories. If this was indeed the ancestral condition, it has been superseded by cooperative competition, presumably because of a favorable cost-benefit analysis. Whether there is a still further derived mutual benefit in the form of synergistic enhancement of territorial maintenance or expansion is an open possibility that remains to be investigated.

9 Ecological Relationships in the Manu Primate Community

Here I provide a qualitative overview of ecological relationships among the thirteen primate species inhabiting the Manu region. Due to the large number of species, it has not yet been possible to complete long-term studies of all of them, and consequently the depth of our knowledge is somewhat uneven. Seven species have been studied intensively through a full annual cycle. These are the five treated here plus *Callicebus moloch* and *Aotus trivirgatus*. The latter two were studied by Patricia Wright in work so recently completed that the results have not yet been published. *Cebuella* was observed for several months in 1977, first by Mariella Leo, and then less systematically by Charles Janson. Katharine Milton undertook a brief set of observations on *Alouatta* in 1975, and these were augmented in 1979 by Carlos Saavedra. What little we know of the ecology of *Lagothrix* comes from a two-month study in 1979 conducted by Marleni Ramirez (1980). Of the common species, the only one remaining is *Ateles*. It is such a conspicuous animal in the study area that we have a great many casual observations of its habits. A long-term project on its feeding ecology is just now getting underway as this goes to press (1982). That makes eleven species for which we have more or less extensive observations, leaving only two, *Callimico* and *Pithecia*, for which the information is wholly inadequate. Fortunately, neither of these is an important member of the community.

Only two synecological studies of New World primates have been undertaken previously. One was by the Hladiks at Barro Colorado Island, Panama (Hladik and Hladik 1969; Hladik et al. 1971). There the community consists of just five species, a far cry from the thirteen of Manu. Recently, Mittermeier and van Roosmalen (1981) reported on the ecology of eight sympatric species in Surinam. A greater number of sites have been investigated in the Old World. Depauperate insular communities have been studied, again by the Hladiks, in Ceylon (Hladik and Hladik 1972) and Madagascar (Hladik 1979; Sussman 1974). Analyses of partial communities have been reported for Cameroon (Struhsaker 1969), Gabon (Gautier-Hion 1978, 1980) and Uganda (Struhsaker and Leland 1979), but an account of all the species in one of these rich African localities has yet to appear. Better success has been achieved in Asia where dietary and behavioral comparisons are available for all sympatric species in two communities: Malaya (Chivers 1973, 1980; MacKinnon 1978) and Borneo (Rodman 1978). The number of species involved, however (six in each case), is only about half of that to be expected in a central African or Amazonian locality. Nevertheless, the data that do exist on Old World primate communities make it clear that they are structured in a very different way from the New World community.

This chapter is organized into three major sections. The first of these examines the implications of body size as it influences the ecological roles that species play in the community. I show that an animal's size not only predisposes it to a certain trophic position, but also is critical in its relationships with other species and in its use of habitat. Species having approximately the same body size are the subject of the second section. Here I inquire into the modes of ecological differentiation that are independent of body size. Then, in the third section, I compare the structure of New and Old World primate communities in an effort to understand some of the remarkably distinct features of several more or less independent adaptive radiations.

Implications of Body Size

Body size is a feature of the greatest importance in structuring New World primate communities (Table 9.1). It carries strong implications for such diverse ecological and behavioral characteristics as diet, resource use, repertoire of foraging techniques, antipredator strategies, locomotion, and habitat selectivity. More directly, body size determines the daily metabolic requirement, and with it, the home range needed to support an individual or group, and ultimately the biomass that can be attained by the species. In the paragraphs that follow I shall discuss the implications of body size in New World primates as deduced from observations on the species at Manu, expanding on the abbreviated statements contained in Table 9.1

Diet. Resources that can be harvested slowly or only in small amounts are the private domain of small-sized consumers. Such resources include sap (gum), nectar, and insects. At Cocha Cashu, sap is the principal year-round energy source for *Cebuella* (100 g), and nectar sustains both *Saguinus* (400–550 g) species through the dry season. The nectar of *Combretum* and *Quararibea* is also avidly consumed by *Cebus* and *Saimiri*, but for them it is better regarded as a supplementary rather than as a primary resource. In the Old World, all obligate and facultative primate gum and nectar feeders are prosimians weighing less than 1 kg (Charles-Dominique 1974; Sussman 1979). Obligate insectivory (exclusive of social insects) is likewise a mode of life

Table 9.1
Ecological correlates of body size in New World primates.

	Small size <1–2kg	*Large size >1–2kg*
Diet	Use of sap and nectar as primary food resources in some species. Mainly soft, ripe fruits eaten. Insectivory partial (omnivory); facultative insectivory possible.	Use of hard/protected plant parts (esp. *Cebus*: nuts, meristems, and immature inflorescences of palms; pith). Unripe fruit, may be eaten facultatively. Insectivory partial or lacking.
Degree of concentration of resources	Specialization on diffuse resources possible.	Use of diffuse resources (mainly arthropods) restricted to at most a minor fraction of the diet.
Foraging techniques	Searching restricted to surfaces or pliable substrates (dead leaves); stalk-and-pounce captures possible.	Destructive searching of protected hiding places (under bark, in hollow branches, etc.) possible.
Locomotion	Cling-and-leap possible; also vine walking and travel through small plants in the understory.	Brachiation possible; also long leaps and catapulting from swaying crowns; firm support necessary for branch walking.
Antipredator strategies	Crypticity, hiding in vine tangles, dense foliage.	Escape in size, i.e., too large or powerful to be handled by most predators in the habitat.
Habitat selectivity	Specialization on patchy habitats frequent; use of vine tangles and understory.	All use forest matrix; lack of firm supports precludes use of vine tangles and small plants of understory.

that is open only to animals of small size, mostly less than 100 g. This is a consequence of the large amount of time necessary to search for prey or to extract them from their hiding places. Omnivores, here defined as mixed feeders dependent on arthropods and other small prey for most or all of their protein intake, tend to be larger than obligate insectivores, but nevertheless appear to be constrained to sizes of less than a few kg. Ten out of the thirteen primates at Manu weigh less than 3 kg, and eight of these (all but *Callicebus* and *Pithecia*) are omnivores. The three large species, *Alouatta*, *Lagothrix*, and *Ateles*, are strict vegetarians.

Being possessed of relatively high metabolic rates, small animals must process food quickly, and hence must feed selectively on high-quality items that are easily digested (e.g., Bell 1970). In contrast, the longer retention times allowed by the slower metabolic rates of larger animals permit detoxification and processing of fiber and other polymerized materials by fermentation and the action of symbionts. This opens up a potentially huge supply of low-value food resources to large species (Kay and Hylander 1978; McNab 1978; Milton 1981; Pough 1973). Thus, folivorous primates are typically large, weighing more than 5 kg. There are only a few exceptions to this rule, the most outstanding being several lemurs (Hladik 1979; Sussman 1977; Jolly 1972), one of which (*Lepilemur*) weighs less than 1 kg, yet attains a biomass (500 kg/km^2) that is elsewhere matched only by much larger species (e.g., *Alouatta*, *Colobus*, *Presbytis*; Charles-Dominique and Hladik 1971). Just why *Lepilemur* and some of its relatives are so much smaller than any folivorous monkey is not known. At Manu, *Callicebus moloch* (ca. 800 g) seems to obtain most of its protein from leaves (25% of feeding time; cf. Chapter 5), but it does not consume them in bulk and its feeding is highly selective (see below).

Large size, through the concomitant increase in jaw and muscular strength, also permits exploitation of hard or protected foods such as seeds, pith, and the nuts and meristems of palms. These constitute significant reserve food supplies for *Cebus apella* in the dry season. Here the importance of size is underscored by the fact that the slightly smaller *Cebus albifrons* exploits these resources to a lesser degree, and then only with obvious difficulty (Chapter 5).

Degree of concentration of resources. Differences in body size allow ecological divergence by means of specialization on resources having distinct levels of concentration. This was pointed out above in the discussion of sap, nectar, and insects as primate resources. All of these exist in low concentration; hence they can be harvested profitably only by an animal with low metabolic demands. The same is true of dispersed and slowly renewing fruit crops of the types exploited by tamarins. Attainable feeding rates are too low to sustain a larger animal. This does not entirely preclude the use of such resources when they are encountered by heavier species, but it does preclude specializing on them. This limits the harvest of dispersed fruits by large monkeys, and leaves a competitive refuge within which the smaller species exist.

This argument does not hold in reverse. Tamarins are as capable of eating the most concentrated fruit resources (e.g., *Ficus*, *Brosimum*) as spider monkeys. In fact *Saimiri*, a relatively small species, is, of all members of the community, most dependent on concentrated resources. That tamarins use such resources very little seems to have more to do with exposure to predators and to possible interference by larger species than it does to the immediate properties of the fruit. Large species must use concentrated resources in order to achieve adequate feeding rates. To the extent that this leads to intensified interspecific competition through aggressive displacements, it could lead to an evolutionary feedback between use of the most concentrated resources and body size. This could be inferred from the existence of an interspecific hierarchy of aggressive control of fig trees in the Kibale forest of Uganda (Waser and Case 1981). Increased size, however, is not the only means of escaping interference competition, as shown by *Saimiri*, which can thwart the aggression of larger species by sheer dint of numbers.

Body size and foraging repertoire. This is a topic that was stressed in Chapter 6 and so will be mentioned here only briefly. Lacking in sufficient strength to tear into insect hiding places other than dead or rolled-up leaves, small insect-hunting primates are restricted to searching exposed surfaces where many of the available prey items are mobile and capable of escaping. Escape-prone prey include many highly desirable large items such as orthoptera, cicadas, lizards, and frogs. Capturing these, except through fortuitous circumstances, requires the stalk-and-pounce technique used by *Cebuella* and *Saguinus imperator*. (To capture this sort of prey, birds typically employ one of two techniques: either they strike while hovering or flying past, or they flush the item and then chase it [e.g., Fitzpatrick 1978].) For a quadripedal animal, approach to within pouncing distance requires stealth and light-footedness, tactics that are out of the question for heavy-bodied animals. For these, brute-force methods can make up for lack of finesse, to a point. By ripping the bark off dead limbs, rummaging through debris, and biting open hollow twigs, dead bamboo canes, and the bases of palm petioles, *Cebus* can extract a steady harvest of ants, termites, and hiding prey of many types. But the fact that such activities are compulsively pursued for most of the day suggests that the rewards are just barely sufficient. Still larger size would not enable access to many more kinds of hiding places, and thus would not be favored in the evolutionary cost-benefit analysis.

Size and escape from predation. A formidable battery of raptors, themselves arrayed over a broad size range, poses a lifelong threat to all but the largest Manu primates (Table 9.2). Generally, the smaller raptor species are more common than the larger ones, and the largest, *Harpia* and *Morphnus*, are highly dispersed, having territories much larger than the home range of any primate. Large raptors can potentially take small prey on occasion, but the converse does not hold. Thus, the smaller primates are exposed to many individuals of several raptor species, while the larger primates need to worry

Table 9.2
Potential primate predators in the raptor community at Cocha Cashu.

Species	*Weight (g)*[a]	*Reported diet*[a]	*Attacks observed at Cocha Cashu*[b]
Harpy Eagle (*Harpia harpyja*)	♂4000–4600 ♀ ?	Sloths, *Cebus*, *Tamandua*, Curassows[c]	*Cebus*, *Alouatta*, *Saimiri*
Guianan Crested Eagle (*Morphnus guianensis*)	♂ ? ♀1750	Monkeys, opossums	*Cebus*, *Saimiri*, *Callicebus*
Black Hawk-eagle (*Spizaetus tyrannus*)	♂ 904 ♀1092–1123	Opossum	
Ornate Hawk-eagle (*Spizaetus ornatus*)	♂ 835–1004 ♀1389–1607	Lg. birds, *Potos*	*Saimiri*, *Saguinus*, *Callicebus*
Black-and-white Hawk-eagle (*Spizastur melanoleucos*)	♂ 750–780 ♀ ?	Birds, mammals	
Slate-colored Hawk (*Leucopternis shistacea*)	♂ ? ♀1000	Snake, frog	*Saimiri*
Bicolored Hawk (*Accipiter bicolor*)	♂ 204–250 ♀ 342–454	Birds	*Saguinus*, *Sciureus*

[a] Weights and dietary information compiled from Brown and Amadon (1968) and Haverschmidt (1968).
[b] From unpublished observations of the author, Charles Janson, Patricia Wright, Richard Kiltie, and others.
[c] Prey species brought to a nest observed by Rettig (1978).

about only a few individuals of one or two predator species. In my opinion, aerial predators constitute the only serious daytime threat to arboreal primates. Ground-foraging *Cebus* or *Saimiri* might occasionally be ambushed by a tayra (*Eira barbara*) or a felid, but we never witnessed such an attack. At night, roosting animals could be seized by stalking felids or large boas, but we have no way of estimating the frequency or importance of such events.

Manu primates employ three distinct strategies to thwart predation, two of them behavioral and the third morphological. These are: (1) crypticity, (2) joining in groups, and (3) escape in size. The last of these applies only to the adults of the largest species: *Ateles*, *Lagothrix*, and, less certainly, *Alouatta*. All the rest employ one of the behavioral strategies, and even between these there is a strong size-dependent dichotomy.

Crypticity is possible only for the smaller species that can move quietly through the dense foliage of the understory and can take refuge within the cramped confines of vine tangles or hollow trunks. All the callitrichids spend many hours per day in locations that are most reasonably interpreted as safe hiding places. *Aotus* does likewise. *Callicebus*, being somewhat too large to move comfortably in vine tangles, spends much of its time sitting quietly underneath umbrella-like canopies of vines or the inner branches of densely foliated midstory trees. *Pithecia* probably does much the same thing, though our observations of it are too few to allow a confident statement. Thus, of the eight species with adult weights of less than 1.5 kg, six or seven evade predators through crypticity and hiding. Only *Saimiri*, which through its predilections for figs and insectivory is obliged to be constantly active, takes the alternative option of seeking safety in large groups. This will be discussed more fully in Chapter 10.

Among the larger species, *Cebus* spp. are too large to hide effectively in tangles, and too active for hiding to be a way of life. Like *Saimiri*, they appear to seek protection in groups. *Alouatta*, *Lagothrix*, and *Ateles* spend a great deal of time at rest, but make no attempt to hide, usually selecting a conspicuous exposed perch in the canopy. From such vantage points they at least may be able to spot an attacking eagle before it is upon them, if indeed they are under any threat at all.

At this point it seems appropriate to digress for a moment to consider the special case of *Aotus* and its singular nocturnal habit. *Aotus* is not merely crepuscular; it is absolutely nocturnal, emerging well after sundown from its traditional daytime roost, and returning to it well before sunup (Wright 1978). By adopting such an unconventional schedule (for a monkey), *Aotus* may avoid aerial predators altogether. The largest owl in most Amazonian forest localities is the spectacled owl (*Pulsatrix perspicillata*, ca. 800 g); it is only questionably large enough to take an adult *Aotus*. Stomachs of *Pulsatrix* examined by collectors have contained mostly insects and small vertebrates (Haverschmidt 1968; J. P. O'Neill, pers. com.).

If *Aotus* has escaped one kind of predation by its nocturnality, it seems to have adjusted to the new circumstances through some other rather unusual habits. In her study of the species in central Peru, Wright (1978) noticed that the groups tended to spend long periods in preferred fruit trees, resting or casually insect foraging between feeding bouts. Working at Cocha Cashu, Emmons confirmed this behavior of *Aotus* at dry season nectar sources (Janson et al. 1981). We point this out as a possibly significant habit, because observers of diurnal frugivores have repeatedly remarked that they do not tend to linger in feeding trees (Klein and Klein 1975; Howe 1979), nor, evidently, do bats (Morrison 1978a, 1978b). What is a poor risk by day might become a good one by night if aerial predators are not a threat. The felids and snakes that could prey on *Aotus* would probably have great difficulty stalking them on branches and thus would be most likely to attack from ambush while the monkeys were moving through the forest. *Aotus* may attempt to minimize its exposure to such attacks by seeking out a particularly large resource tree and remaining in it for long periods.

Many readers may be wondering what more direct evidence we have of predation in primate populations at Cocha Cashu. There is some, as will be described below, but it is necessarily anecdotal, because successful predation is a rare event—at most it can occur only once in the lifetime of a prey. At the level of a whole troop it is also a rare event. A *Cebus apella* group, for example, could not sustain more than one loss to predation per year without going into decline, because the birth rate is only 1–2 per year. From this perspective I feel that predation is an important source of mortality, and perhaps the leading source of mortality in the population we studied.

From the reasoning of the previous paragraph, one could expect to witness an average of about one successful predation in a full year of watching primates. We did in fact witness one predation event: a juvenile *Saguinus fuscicollis* was killed during the April sample by a small hawk which we took to be *Accipiter bicolor*. On another occasion, an ornate hawk eagle (*Spizaetus ornatus*) was seen carrying a full grown *Saguinus imperator*. The individual was not from our study troop, however. In addition, there were at least three further instances of presumptive predation. Of the four young born into our two *Saguinus* study troops, three had disappeared before they were six months old, including the one we saw killed. During May 1977, the dominant male of our *Cebus apella* study troop suddenly disappeared one afternoon while we were following the group. A few days later the skull was found very near to where the animal had last been seen alive. Although we did not see what happened in this case, a healthy adult *Cebus* is not likely to drop dead spontaneously.

Unsuccessful raptor attacks were a relatively common occurrence in our experience, with one or more being recorded in nearly all the three-week samples. They were relatively more frequent on *Saimiri* and *Saguinus* (usually by *Spizaetus ornatus*) than on *Cebus* (only by *Harpia*; two attacks). It is thus impossible to deny the existence of predation on Neotropical primate populations, although one could still argue about its importance as a selective force. Our strong impression from firsthand experience is that it is very important.

Size and locomotion. All the callitrichids at Manu more or less routinely engage in cling-and-leap locomotion, though they differ in how much they use it. *Saguinus fuscicollis* is a habitual cling-and-leaper, employing this locomotion almost exclusively in its insect-hunting activities (Chapter 6). *Cebuella*, the pygmy marmoset, similarly spends much of its time on vertical surfaces as it opens and visits the pits it exploits for sap. *Callimico* moves through the dense understory of the swamp forest it inhabits by springing from stem to stem, but we have no observations on the kinds of locomotion it uses while foraging. Of the four callitrichids, *Saguinus imperator* is the least prone to trunk hopping, usually preferring to make its way along branches and vines. The violent forces of acceleration and deceleration involved in cling-and-leap locomotion, and its requirement for fail-safe traction on smooth trunks, would seem to preclude its use by all but the smallest primates. The remarkable Indrids of Madagascar (to 12.5 kg), however, stand out as a glaring and enigmatic exception to this rule (Jolly 1972).

While cling–and–leaping is an option for squirrel-sized creatures, brachiation can be employed successfully only by large ones, because of the long lever arms needed to generate high-velocity swings and to clear the terminal foliage of branches. Among the New World primates, brachiation is employed only by *Ateles*, *Brachateles*, and *Lagothrix*, all of which weigh in excess of 5 kg.

Size and habitat selection. I have already intimated in the discussion of locomotion some of the ways in which body size controls habitat use. Certain microenvironments within the forest are constructed of plant parts that are too flexible or feeble to support the weight of a large animal. These include bamboo thickets, canopy vine tangles, and the slender shrubs and trees of the understory. Such places thereby become competitive refuges for the smaller animals of the forest, offering both shelter and resources that are unavailable to larger species. It is perhaps not surprising then to note that of the eight Manu primates weighing less than 1.5 kg, six are habitat specialists (only *Aotus* and *Saimiri* are not). On the other hand, none of the five species weighing more than 1.5 kg shows similar tendencies. (Our operational definition of a habitat specialist here is a species that is not distributed through high ground forest, the habitat that covers more than 90% of the Manu region.)

Cebuella occupies edge situations along river and stream banks where there are festooning curtains of vines to shield its activities (see also Ramirez et al. 1977; Castro and Soini 1977). We have found *Callimico* only in a type of swamp forest having a thick understory composed of many small-diameter trunks. This is similar to the habitat that has been described for the species in Bolivia (Pook and Pook 1981). The two *Saguinus* species are more generally distributed, but their territories always contain substantial habitat diversity, including abundant edge, either along the margins of swamps, lakes, streams, etc., or in the form of numerous treefall openings. *Callicebus moloch* lives in the same sorts of places. Within expansive tracts of high-canopy forest, one encounters neither *Saguinus* nor *Callicebus*. *Pithecia* also has some restricting habitat requirement within the region, but our experience with it is too limited to permit a diagnosis. *Saimiri* is not a habitat specialist; it wanders everywhere, though its use of distinct vegetation types varies seasonally (cf. Chapter 7). About *Aotus* we know rather little except that several groups are distributed around the study area, and that these move freely through several habitats, including high-canopy forest. What the habitat specialists have in common is a liking for edges, the understory, and heavy concentrations of vines, situations in which they have an advantage due to their small size.

Turning now to the larger primates, we can surmise that their greater metabolic demands would preclude specialization on patchy habitats. Individual patches would not be big enough to support a group, and much time and energy would be wasted in moving between patches. It would be energetically more efficient to use the predominant habitat, the matrix within which the minor habitats are imbedded, and to exploit the latter opportunistically as convenient. This is indeed what the larger primates all seem to do. Even *Lagothrix*, where present, appears to follow this rule.

Here it is critical to make clear the functional distinction between the use of patchy habitats and patchy resources. An animal that specializes on a patchy habitat will be restricted to locations where that habitat occurs. In contrast, an animal that specializes on patchy resources will be highly mobile and relatively catholic in its use of habitats, requiring mainly the (temporary) availability of concentrated food resources. A large-bodied species may very well specialize on patchy resources, but it is unlikely that such an animal would be restricted to living in a narrowly defined habitat, particularly one that was highly fragmented.

Within any habitat, however, large animals may be restricted by their size to certain microenvironments. A requirement for sturdy supports thus obliges the heavier arboreal primates to concentrate their activities in the middle and upper strata (Fleagle 1980). Here they can dominate the available fruit resources, both because they can easily intimidate or expel smaller species, and because the latter are in any case reluctant to expose themselves in such open situations. Among the smaller species, only *Saimiri*, which derives protection from its own numbers and from its association with *Cebus*, is a full member of the canopy community. But even *Saimiri* and its *Cebus* mentors are reluctant to linger in the wholly exposed crowns of emergents. These are the special domain of *Ateles*, *Lagothrix*, and *Alouatta*.

Ecological Relationships of Like-Sized Species

So far in this chapter I have considered how primates that differ in body size face a distinct set of evolutionary opportunities and constraints that necessarily channel them into divergent roles in the economy of the forest. Now we shall undertake an examination of the modes of ecological divergence that serve to differentiate the roles of species that are similar in size.

The callitrichids (100–600 g). Cebuella is ecologically isolated by virtue of its highly specialized feeding on sap. It experiences periodic harassment from *Saguinus fuscicollis* in the form of piratical raids on its sap trees, but these are probably of minor consequence. Better than any other species it can move without shaking branches, a fact that it exploits by stalking mobile prey at an almost imperceptible pace before it strikes with lightning speed.

The two *Saguinus* species differ in their locomotory behavior, a fact that allows them an almost perfect spatial separation of their insect-hunting activities. As joint proprietors of a common territory, they share a single pool of fruit resources which they use in apparently identical fashion. The location of the territories in habitats having high densities of resource trees, and the intensified defense of them during periods of resource scarcity, both seem to suggest that fruit is the critical limiting resource. Thus, it is paradoxical that their only apparent mode of ecological segregation relates to their prey-hunting activities.

It is hard to imagine that prey are really limiting, or that one species would seriously depress the abundance of prey available to the other even if they foraged in the same way. At most, the groups search only a few percent of the appropriate substrate in their territories each day. Moreover, their use of space within the territories is concentrated in a small area near the center, whereas a more even coverage would be expected if prey depletion were a major factor in regulating their movements. In any case, compared to the much larger biomass of birds and insect-hunting primates of other species, their impact on prey populations could not be great. The situation is puzzling and I shall have to leave it unresolved.

As for *Callimico*, we know next to nothing about it. We have registered just two encounters, both of small (family?) groups traveling low (<4 m) in an area of swamp forest that is also occupied by both species of *Saguinus*.

Callicebus, Aotus, Saimiri, and Pithecia (0.7–1.4 kg). The members of these disparate genera are similar in weight, but the resemblance ends there.

Callicebus is a vegetarian, deriving its protein intake largely from leaves. Many of the leaves it consumes are from the growing tips of vine runners. The only other partially folivorous primate in the community, *Alouatta,* is effectively excluded from this resource because vine shoots are highly dispersed and cannot be harvested quickly or in large quantity in any location, and because the outer surfaces of vine mats would provide little support for so large an animal. *Callicebus* also differs from the other species in its size class in consuming unripe fruit, especially figs. The combination of unripe fruits and vine leaves is unique and assures the ecological isolation of *Callicebus*. Both of these resources are likely to be more evenly distributed in space and time than ripe fruit. Moreover, the ripe fruits eaten by *Callicebus* include many of the species used by *Saguinus*, and these as well constitute a relatively stable and diffusely distributed resource. Perhaps these circumstances then account for the avid territoriality of *Callicebus*.

From our few observations on the feeding of *Aotus*, we note that they take fruits that are highly preferred by *Cebus* and *Saimiri*. By visiting the trees at night, *Aotus* certainly avoids interference competition, and may lessen its exposure to exploitation competition as well. This is possible to the extent that individual fruit crops ripen continuously, day and night, over relatively contracted periods. When this happens, a tree that has been thoroughly harvested in the afternoon by diurnal primates may again contain ripe fruit several hours later. During the early dry season when soft fruit is in short supply, *Aotus* switches to nectar, putting it in competition with *Saguinus, Cebus*, and *Saimiri* (Janson et al. 1981). Here again nocturnality may be advantageous as the flowers in question (*Combretum* and *Quararibea*) produce nectar on a 24-hour schedule (our unpublished results).

The relationship of *Saimiri* to its closest ecological neighbors, *Cebus apella* and *albifrons*, has been extensively documented in previous chapters. It broadly overlaps *Cebus* in its choice of fruits, but makes only casual inroads on the main crops used by *Saguinus* and *Callicebus*. In its insect-foraging habits, it is also more akin to *Cebus* than to *Saguinus*, especially to *C. albifrons* (Chapter 6). When hardest pressed in the early dry season before the first nectar becomes available, it resorts to total insectivory, sometimes for days at a stretch. It is during these periods that the selective advantage of being small is obvious. A larger animal simply could not subsist on the insects it could catch in a 12-hour day. To *Saimiri*, facultative insectivory is the recourse of last resort in coping with the exigencies of a seasonal environment and in permitting coexistence with *Cebus*.

We now come to *Pithecia* which, like *Callimico*, is little known in the wild. It appears to be restricted in habitat at Cocha Cashu, as judged from the fact that all our encounters have been in one sector of the study area. However, the vegetation in this area is a mosaic of streams, swamps, and high forest, so no inferences about its requirements could be justified without further study.

Cebus apella and Cebus albifrons (2–4 kg). The subtle and interesting ecological distinction between these two congeners is a major topic of discussion in several sections of the book and will not be extensively reiterated here. The main difference is in their ranging behavior and in the various consequences that follow from it. In addition there are quantitative differences in their insect-foraging behavior (Chapter 6) and in their use of certain plant materials (Chapter 5): pith and palm products (more by *apella*); figs and wasp nests (more by *albifrons*). In most respects the species are very similar and are clearly each other's ecological nearest neighbors. Divergence in their diets and foraging behaviors is most marked during the annual period of scarcity, a trend that has been noted as well for sympatric species of *Cercopithecus* in Gabon (Gautier-Hion and Gautier 1979). As for their relationships to species in adjacent size classes, the *Cebus* species overlap most with *Saimiri* (figs and other soft fruits; insects) and with *Ateles* (many fruits, e.g., *Ficus, Brosimum, Inga, Cecropia, Strychnos*, etc.).

Ateles, Lagothrix, and Alouatta (5–10 kg). These three species constitute a natural ecological grouping. The diets of all of them consist principally of fleshy fruits, and within the list of fruits eaten there is considerable overlap. It may be that the important distinctions between them lie in their means for acquiring an adequate protein intake, though at present this is largely conjectural.

Alouatta is the most folivorous of the New World monkeys, and at certain seasons leaves may constitute as much as 60% of its diet (Hladik and Hladik 1969; Milton 1978). Most of the leaves are eaten during expansion ("flush leaves"). Indeed, flush leaves constitute the most distinctive element of the *Alouatta* diet, a fact we can confirm from casual observations on *A. seniculus* at Cocha Cashu. The fruit component of its diet is quite ordinary, consisting of a variety of soft fleshy species that are eaten avidly by other monkeys, particularly *Cebus* and *Ateles* (e.g., *Ficus* spp., *Brosimum, Pseudolmedia*, etc.). The main difference is that *Alouatta* seems to be able to eat many of them before they ripen.

The social and ranging behavior of *Alouatta* differ markedly from those of the other two species. It lives in small, stable, single-male groups which occupy relatively small home ranges of 20–30 ha. In gallery forest in the Venezuelan llanos, where population densities of *A. seniculus* are several times as high as at Cocha Cashu (up to 150 vs. 30 per km^2), the home ranges are smaller and overlap extensively (Rudran 1979). The ability of such a large animal to live in so small a home range presumably derives from its dependence on abundant resources, and from the fact that there is a considerable seasonal complementarity in the availabilities of fruits (rainy season) and flush leaves (peak in late dry season) (Leigh and Smythe 1978).

Both *Ateles* (Klein and Klein 1975) and *Lagothrix* (Ramirez 1980) live in fission-fusion societies, the members or subgroups of which travel widely within home ranges encompassing at least a few km^2.

Lagothrix has a peculiar distribution in Peru, being found here and there in the lowlands but being absent over large areas (Freese 1975). Its absence in some cases may be attributable to overhunting, for it is the most favored of all the primates as game; but this can hardly be the explanation for its spotty occurrence in the Manu region. According to the reports of numerous local residents, it is found only on the right (south) bank of the river. It is absent in the vicinity of Cocha Cashu and from the mouth of the river far downstream. In between, it is common near the park guard post at Pakitza, about 30 km downstream, and along the Rio Pinquen, a major right-bank tributary. Upstream of Cashu, it is absent for about 60 km until one reaches the level of the Rio Sotileja, where it is again reported as being common by the local Machiguenga Indians. What is to account for the large hiatuses in its population is completely mysterious. We have seen it near Pakitza where Ramirez (1980) carried out her study in the summer of 1979. The observations of Ramirez suggest nothing remarkable about its foraging or dietary habits. During August it ate the same common fruits (*Ficus, Brosimum*) that were being consumed by the other large primates, supplemented with a minor component of leaves. Groups, numbering about 15 individuals, frequently fragmented and re-formed, though the subgroups were generally larger than is the case in *Ateles*. The animals roamed widely within a home range of 2.5–4 km^2 (Ramirez 1980).

Just how *Lagothrix* differs from *Ateles* in diet, group organization, and ranging behavior is not yet clear. The two seem to be very close in all these aspects of their ecology. Indeed, they appear sufficiently alike to raise the possibility that it is competition from *Ateles*, or the combination of *Ateles* and *Alouatta*, that is to account for the mosaic distribution of *Lagothrix*. Perhaps it can enter the community only where one of the other two is absent (e.g., many mountainous regions above the altitudinal limit of *Alouatta*), or in particularly productive sites (e.g., volcanic or alluvial soils [cf. Eisenberg 1979]), or where certain critical tree species are present in the forest. This is an interesting issue that deserves further attention in the future.

Of the three species, *Ateles* is clearly the most frugivorous. Klein and Klein (1977) found in Colombia that ripe fruit made up 83% of the diet, a figure that is consistent with our more casual observations. Leaves and buds accounted for only 5% of feeding time, unripe fruit for less than 1% and animal prey for 0%. At Cocha Cashu, *Ateles* often feeds on the fruits of various lauraceous trees. These are green in color (though many have showy receptacles) and extremely bitter to the taste. They are ignored by most of the other monkeys (*Callicebus* eats them occasionally), perhaps because they are difficult to digest. In Colombia, palms (*Iriartia ventricosa* and *Euterpe* sp.) were the principal source of nutrition (53% of feeding time) during the period of scarcity of soft fruits (Klein and Klein 1977).

It is interesting to note that the palm genera used by *Ateles* are different from the ones concurrently exploited by *Cebus* (*Scheelea, Astrocaryum*). The fruits of *Iriartia* and *Euterpe* are similar in being constructed rather like dates, except with a much thinner fleshy coating (pericarp) surrounding a relatively larger seed. The amount of reward per fruit (only the pericarp is eaten) is very small. Although *Cebus apella* occasionally eats *Iriartia* at Cocha Cashu, and is reported by the Kleins to use *Euterpe* in Colombia, it strongly prefers *Astrocaryum* and *Scheelea*, both of which offer far greater nutritional rewards (cf. Chapter 5). The nuts of both species require great force to break open, a fact that may exclude them from the diet of *Ateles* (Izawa and Mizuno 1977; Struhsaker and Leland 1977; Kiltie 1980). Thus, the dry-season food supply of *Ateles* seems to be well differentiated from that of *Cebus*, except for figs which are eaten opportunistically whenever they are available. *Alouatta*, with its predilection for flush leaves and unripe fruit, also has a distinctive dry-season diet. We have not observed *Alouatta* using palms of any kind, nor have they been reported to use them in Colombia (Klein and Klein 1975) or Panama (Hladik and Hladik 1969).

Putting *Lagothrix* aside, for we know so little about it, we can conclude that all the larger primates at Cocha Cashu differ in one or more major components of their diets; and, as we have repeatedly noted, dietary divergence between the species seems most pronounced during the annual period of resource scarcity. By extension, it is probably safe to say that this will prove to be true also of *Callimico*, *Pithecia*, and *Lagothrix* as soon as adequate information on these species becomes available.

With only a limited sample of species from which to draw inferences, it is not possible to arrive at any far-reaching generalizations about the types of adaptations that characterize the ecological distinctions between like-sized species. Nevertheless, there is one consistent pattern that holds for all the sets of species. In every case there are differences in diet—not so much absolute differences as differences in the proportions in which certain resources are consumed. We see this among callitrichids, with *Cebuella* depending on sap as the mainstay of its subsistence, and *Saguinus* spp. using it casually as opportunity permits; in the greatly differing emphasis on ripe fruits and arthropods shown by *Callicebus* and *Saimiri*; in the contrasting ratios in which figs and palm nuts are consumed by the two *Cebus* species; and in the importance leaves assume in the diets of *Ateles*, *Lagothrix*, and *Alouatta*. In addition, there are other ways in which the species diverge—in foraging technique (*Saguinus* spp.), in nocturnal vs. diurnal habits (*Aotus* vs. *Callicebus* and *Saimiri*), and in ranging patterns (*Cebus* spp.; *Ateles* vs. *Alouatta*). Thus, while diet is in general correlated with body size (decreasing insectivory and increasing folivory in larger species), the correlation is a loose one that has left open adaptive opportunities for the diversification of species at each size level.

Ecological Contrasts with Old World Primate Communities

Our analysis of ecological relationships among New World primates has stressed the critical importance of differences in body size as the primary evolutionary organizing principle of the community. Adult weights scatter over nearly a 100-fold range. This not true in the Old World. The maximum size of New and Old World arboreal monkeys is about the same (ca. 10 kg); it is the minimum size that differs so strikingly, there being only one Old World monkey that weighs less than 4 kg—the talapoin (*Miopithecus [Cercopithecus] talapoin*: 1–1.5 kg), a species with a restricted range in the Congo Basin. In looking at the structure of whole communities, one is more impressed by the differences between the New and Old World situations than by the similarities.

Take, for example, the Malayan forest primates studied by Chivers (1973, 1980) and his colleagues. Six species of monkeys and apes inhabit their study area at Kuala Lompat: langurs (*Presbytis obscurus* and *P. melanophos*), two macaques (*Macaca fasicularis* and *M. nemestrina*), and two hylobatid apes (*Hylobates lar* and *Symphalangus [Hylobates] syndactylus*). The adult weights of all of these fall within a factor of three (4–11 kg). Three of the species devote somewhat more than 50% of their feeding time to eating leaves (the two *Presbytis* species and *Symphalangus*), while in the other three, leaves account for somewhat less than 50% and fruit for most of the rest. Dietary distinctions among the species are thus far more subtle than among the New World primates. Differences are primarily quantitative, appearing in the rank orders in which different species of plants feature in the diets (MacKinnon and MacKinnon 1980). Other resources that enter into the diets of New World species, such as sap, nectar, pith, and hard nuts, are not used at all. Furthermore, in none of the Malayan species did animal matter make up more than 15% of the feeding observations. However, vertical stratification in the forest is conspicuously more important in separating feeding activities than in the New World community. The contrasts between the two situations are thus many and profound, and it is not easy to imagine what is at the root of them. Perhaps in the more equitable climate of Malaya, leaves, especially young leaves, are a more abundant and seasonally reliable resource. This might explain the strong folivorous tendencies of all the Malayan species, as well as their large average size.

The local diversity of primate species in African forests is considerably greater than in Asia. As many as five species of *Cercopithecus* can occur together at a single site (Gartlan and Struhsaker 1972; Gautier-Hion and Gautier 1974, 1979). In addition, one can expect two species of colobus (*Colobus* spp.), one or two mangabeys (*Cercocebus* spp.), a large terrestrial drill (*Mandrillus* sp.), and one or two large apes (*Pan* and *Gorilla*). All these add up to a total of 10–12 species in the richest localities, not including nocturnal prosimians, of which as many as five may inhabit the same forest (Charles-Dominique 1974).

As is true in Asia, African monkeys are decidedly larger than their South American counterparts. The various *Cercopithecus* species range from about the size of *Cebus* (ca. 3 kg) up to 6–7 kg. Mangabeys are larger, comparable to *Ateles* or *Alouatta* (males 9–10 kg). Only the talapoin weighs less than 3 kg. In bulk (1–1.5 kg), diet, ranging habits, and group size (to 60 or more) it is strikingly reminiscent of *Saimiri* (Gautier-Hion 1970, 1971, 1973), and like *Saimiri* it devotes a great deal of time to foraging, enough that animal matter comprises up to 50% of stomach contents (Gautier-Hion 1971, 1973).

Though systematic studies of sympatric *Cercopithecus* species are as yet few in number and tend to be sketchy in detail, ecological distinctions between the species appear to be based on multiple criteria. Ranging patterns differ, as do quantitative aspects of the diet and vertical use of the forest (Struhsaker 1969; Gartlan and Struhsaker 1972; Gautier-Hion and Gautier 1979; Rudran 1978; Struhsaker and Leland 1979). In addition, certain pairs of species seem to segregate wholly or partially by habitat, a finding that applies to mangabeys as well (Gartlan and Struhsaker 1972; Gautier-Hion et al. 1980). As among the Asian primates, size differences have not been regarded as especially important, except in the context of interspecific aggression for control of fruit trees (Waser and Case 1981).

Beyond the example of *Miopithecus* and *Saimiri*, it is not easy to identify cases of convergence among the forest monkeys of Africa and South America. Nevertheless the prosimians, particularly the galagos, do seem to have quite close counterparts in the marmosets and tamarins. Not only do the two groups overlap broadly in size, but their diets are similar as well, with varied emphasis in different species on gums, nectar, insects, and dispersed fruit sources (Charles-Dominique 1974). Although the galagos are nocturnal and the callitrichids diurnal, the convergence in their respective modes of life is about as close as could be expected in two entirely separate adaptive radiations.

At the most general level of organization, the African and Asian primate communities are much closer to one another than either is to the South American community. This is seen in their greater biomasses dominated by folivore-frugivores and frugivore-folivores (Table 9.3). More than three-fourths of the biomass of the African and Asian communities is made up of species in these categories, while at Cocha Cashu such species (*Alouatta* and *Callicebus*) comprise less than one-third of the biomass. Looking only at frugivores and omnivores, the Asian and South American totals fall below the value for Africa. African forests are also particularly rich in artiodactyls and other large terrestrial herbivores (Dubost 1979). Although based on data from only one

Table 9.3

Biomass by trophic class in some New and Old World primate communities (kg/km^2).

	Folivore-frugivore[a]	*Frugivore-folivore*	*Frugivore*	*Frugivore-folivore-insectivore*	*Frugivore-insectivore*	*Gum-insects*	*Total biomass*
NEW WORLD LOCALITIES							
Cocha Cashu, Peru		200	175		275	1	651
Barro Colorado Is., Panama[b]		365	5		50		420
OLD WORLD LOCALITIES							
Kibale, Uganda[c]	2,010		185		330		2,525
Morondava, Malagasy Rep.[d]	2,200	400			80	40	2,720
Polonnaruwa, Sri Lanka[e]	1,500	1,200		250	25		2,975
Kuala Lompat, Malaysia[f]	220	600		250			1,070
Kutai Reserve, Kalimantan[g]	120	80	70	200			470

[a] Classes based on feeding times: Folivore-frugivore if >50% leaves; frugivore-folivore if >50% fruit; frugivore if <10% leaves, etc.; species assigned to one of the insectivorous classes if >10% insects.

[b] From Eisenberg and Thorington 1973.

[c] From Struhsaker and Leland 1979. The data represent five common species; the community includes two additional uncommon species, *Cercopithecus l'hoesti* and *Pan troglodytes*, for which data are lacking.

[d] From Hladik 1979.

[e] From Hladik 1975.

[f] From Chivers 1973 and Clutton-Brock and Harvey 1977a.

[g] From Rodman 1978.

more or less undisturbed locality per continent, the contrasts are great enough to be convincing. This is especially true of the enormous difference in folivory between the New and Old World sites. On Barro Colorado Island, Panama, the deficiency in folivorous primates is partially compensated by an abundance of sloths (Eisenberg and Thorington 1973; Montgomery and Sunquist 1975); but there is no such compensation at Cocha Cashu, where sloths are extremely rare (<1 sighting per man-year in the field).

How can one account for the intercontinental contrasts? There are possibilities of a historical-evolutionary nature as well as of a contemporary ecological nature. Neotropical vegetation may be chemically better protected against folivory. Or perhaps there has not been enough time for specialized primate folivores to evolve in the New World. On the ecological side, it could be that seasonal fluctuations in fruiting and leafing are typically greater in the Neotropics. This is the explanation I tend to favor. If true, it could explain the lower overall primate biomass, because the carrying capacity of the environment is established by resource levels at the worst time of year. Inasmuch as ripe fruit and flush leaves seem virtually to disappear from the Neotropical forest for two to three months a year, a lower biomass of arboreal and terrestrial mammals could be expected. Moreover, during the annual period of scarcity, insects and other small prey may constitute one of the few reliably available resources. If so, this circumstance could at once account for the small mean size of the New World primates, and consequently their greater size range, as well as their higher trophic status and reduced biomass vis-à-vis their Old World counterparts. But until better quantitative data on resource phenology become available for the different tropical regions of the world, one can do no more than speculate on the intriguing differences in primate community structure.

Summary

Ecological divergence among the thirteen coexisting primate species in the Manu region is primarily based on differences in body size. Differences in size over the range represented (0.1–10 kg) carry profound implications for several important aspects of primate existence: diet, exposure to predation and its countermeasures, repertoire of foraging techniques, and habitat utilization in both the horizontal and vertical dimensions. Ranging and territorial behavior are probably also constrained by size, though only indirectly (Milton and May 1976).

Small primates (<1.5 kg) can engage in a number of life styles that are physically limiting or economically unrewarding for larger species. These include sap and nectar feeding, facultative insectivory, hunting by stealth, cling-and-leap locomotion, use of the understory and/or vine tangles, crypticity as an antipredator strategy, habitat specialization, and systematic exploitation of dispersed resources, including piecemeal ripening fruit crops. In turn, large primates (>1.5 kg), by virtue of their greater strength, lower relative metabolic rates, and reduced susceptibility to predation, have open to them an array of options that are closed to smaller species. Among these are the use of leaves and/or unripe fruit as protein sources; use of protected plant parts such as pith and the nuts, meristems, and immature inflorescences of palms; robust foraging techniques such as stripping bark and splitting open bamboo and hollow stems; and extensive use of the exposed high canopy. All these differences comprise the multidimensional framework of resource partitioning and niche differentiation among Cocha Cashu primates.

When one examines the ecological relationships of species occupying the same size class, still further modes of resource partitioning become apparent. Within the marmosets and tamarins, one is a specialized sap feeder (*Cebuella*), while the two *Saguinus* species differ markedly in the substrates searched during insect foraging. The ecology of the fourth member of this group, *Callimico*, remains unstudied. The next size class (ca. 0.75–1.5 kg) contains four species; one is nocturnal (*Aotus*); another is a vegetarian that obtains protein by harvesting vine and bamboo leaves (*Callicebus moloch*); a third is a fig specialist which, in time of stress, can become a facultative insectivore (*Saimiri*); and the fourth, (*Pithecia*), consumes large quantities of seeds (Mittermeier and van Roosmalen 1981). Although nearly indistinguishable in their ecological behavior during periods of fruit abundance, the two *Cebus* species diverge in diet, in the representation of substrates searched during insect foraging, and in ranging behavior for the months when fruit is scarce. *Cebus apella* concentrates its efforts on palms, both for insect foraging and for nuts, pith, and meristems. *C. albifrons* becomes more nomadic, searching widely for scattered ripe fig trees, and more terrestrial in hunting insects and other prey. Finally, the three largest species, all vegetarians and residents of the high canopy, differ appreciably from one another in the proportion of leaves in the diet. Moreover, at least two of them, *Alouatta* and *Ateles*, exhibit dietary divergence during the dry season. *Alouatta* increases its intake of flush leaves and unripe fruit, while *Ateles* exploits the fruit of two genera of palms that are largely ignored by other species. *Lagothrix*, though intermediate between the last two species in its leaf-eating tendencies, remains unknown, both as to the distinctive components of its dry season diet and as to the cause of its singular mosaic distribution.

Members of Old World primate communities are much less divergent in size. Vertical stratification within the forest structure is identified as an important mode of ecological differentiation, especially among frugivorous species. African and Asian primate communities tend to be strongly dominated by folivores and to attain substantially higher biomass densities than their New World counterparts. Wider seasonal fluctuation in resource levels, especially of ripe fruit and flush leaves, is proposed as a possible cause for the smaller average size and generally higher trophic status of New World primates.

10 Synthesis and Conclusions

In this chapter I shall try to resolve some major issues that have been lurking just below the surface in several of the preceding chapters. These issues concern the adaptive interrelationships between home range size, strength of territorial behavior, and group size. Specifically, I shall begin by considering how the distribution of resources in the environment controls the home range size of consumer organisms, and how the varying defensibility of resources leads to a wide range of behavioral responses when conspecific groups meet. I shall then examine the correlation between group size and home range size in order to reach a conclusion about which is the independent variable. Having done this, I shall continue with an analysis of the optimization of group size. The chapter closes with some concrete suggestions about how the environment selects for the several types of social systems that are observed in primates. But before the discussion opens, a couple of comments on procedure are in order.

One concerns the inferential and intuitive methodology used to arrive at conclusions. Many of the ideas and preconceptions we held at the outset of this study were subsequently proven incorrect or irrelevant. I did not proceed from hypothesis to tests of the hypothesis, but rather from one discovery or realization to another. The regimen of observations and measurements that we followed for nearly fourteen months was designed to gather what we thought was relevant information, but much of it was, basically, first-order descriptive natural history. Most of the science entered later, months later, toward the end of our year in the field when we had begun to appreciate the profound impact that the seasonal environment exerts on the lives of our study animals. So, in a procedural sense, we put the cart before the horse in gathering the data before the hypotheses were formulated. Looking back now, it is hard to see how we could have done otherwise without closing our eyes to a lot of things that at first we did not notice or understand, but which later became clear when fitted into a larger context.

The second comment concerns the circularity that often lurks in deductive evolutionary arguments. Lacking any direct way of reconstructing history, we can reason only from what we see, and what we see are the results of evolution, not its causes. It thus becomes difficult to disentangle causes from effects. Did squirrel monkeys, for example, evolve their present ranging behavior to exploit huge fig trees, or did huge figs evolve to exploit squirrel monkeys? Even though it seems more reasonable that squirrel monkeys evolved to exploit figs, because figs are apparently much more vital to squirrel monkeys than vice versa, there is no way of knowing this with certainty. Thus, it cannot be assumed that the selective forces that maintain an adaptation in its present state are necessarily identical to the ones that were operative in its evolutionary origin. Since the earliest origins of behavioral adaptations are beyond our empirical reach, we are obliged to concentrate on analyzing the forces that stabilize them in the contemporary milieu. Although the mechanisms so deduced may be considered proximal ones, it makes them no less interesting or valid as subjects of scientific inquiry.

Ecological Correlates of Territoriality

What is the Controlling Resource?

I shall follow conventional wisdom here and assume that the various manifestations of territorial behavior (or lack thereof) shown by the animals reflect a varied set of costs and benefits involved in gaining privileged access to certain units of space and the resources contained within them (Brown 1964; Brown and Orians 1970). In order to delve into this any further, the important resources must be identified. The most likely candidates, ones that have been proposed for other primate species, are safe sites, insects, and fruit (Hladik 1975; Mitani and Rodman 1979). In our case, all the species in question wander freely within their home ranges and spend the night wherever they are, or in one of a number of traditional roost sites that are scattered throughout the territory. This seems to rule out the availability of safe sites as a limiting resource. Insects are a more serious proposition. Hunting them is the most time-consuming activity engaged in by any of the species, and they are a major source of protein, if not the major source. Yet it seems unlikely that the quest for insects has a decisive influence on the spatial utilization patterns of any of the species.

The arguments for this view were laid out in full in Chapter 7 and need only to be summarized here. A relatively uniform distribution of prey can be inferred from the slow and methodical progress the animals make while foraging. Efficient harvesting of a slowly renewing, diffusely distributed resource would call for a uniform spatial utilization pattern. Yet the patterns shown by all species were conspicuously nonuniform. Large portions of the home range of every species were neglected or underutilized in every sample, while others, notably those near fruit trees, were overutilized. Moreover, seasonal changes in home range use were clearly associated with shifting patterns of fruit availability. The evidence, though it is all circumstantial, points strongly to fruit, not prey, as the ordering constraint on the spatial utilization patterns of all five species, and for the sake of further argument, it shall be assumed that this is the case.

Site Fidelity and Home Range Size

I pointed out in Chapter 7 and elsewhere that the species studied form a graded series with respect to their use of space. The *Saguinus* live in relatively small, unchanging, precisely bounded, and vigorously defended territories. *Cebus apella* is much less rigidly constrained in where it goes. Like the tamarins, it shows a high degree of site fidelity to the central portion of its home range, but the intensively used core area, as well as the total area covered, expands and contracts over more than a twofold range in accordance with seasonal fluctuations in fruit abundance. *Cebus albifrons* goes a long step further toward opportunistic nomadism. The central portion of its home range, as well as the limits, are ill defined by recurrent usage. The troops seek out exceptional concentrations of fruit and exploit them to exhaustion. Their use of different habitats is markedly seasonal and shifts to take advantage of staggered fruiting peaks. All these trends are even more pronounced in *Saimiri*.

It seems evident that both site fidelity and ranging area are controlled by the degree of patchiness of the resources being harvested. Resources, such as those used by the tamarins, which are relatively common, small, evenly distributed in space, and display a high continuity in time, can be most efficiently exploited by an animal that knows where they are and that moves in an orderly fashion from one resource to the next. At the opposite extreme is a species like *Saimiri* which looks for rare resource bonanzas, and which must travel great distances to reach them. Once a bonanza has been reaped, there is no call to return to the site in less than the renewal time, many months or a year later. Site fidelity could only be counterproductive. Adaptive specializations for exploiting resources that are produced with varying degrees of spatial and temporal continuity will entail distinct and characteristic modes of spatial utilization. Continuity of resources favors sedentariness, territoriality, and an intimate familiarity with every recess of the homestead. On the other hand, discontinuity of resources in space and time favors nomadism or migration, opportunism and lack of possessiveness.

Are there other possible explanations for the observed gradient of behavior? Site fidelity might be expected to scale inversely with home range area for reasons that are independent of resource distribution. When the diameter of the home range becomes larger than the distance traveled per day, a uniform or graded usage pattern is no longer possible on a day-by-day basis, although such a pattern may be realized when movements are integrated over longer time periods (Mitani and Rodman 1979). A requirement for large amounts of living space, however, does not automatically preclude territoriality. Many top predators—tigers, pumas, eagles, and wolves, for example—advertise and defend territories that are sometimes tens of square kilometers in extent (Hornocker 1969; Schaller and Crawshaw 1980; Sunquist 1981). Thus, a high degree of site fidelity is not necessarily incompatible with the occupancy of a large home range. So long as resources are uniformly distributed, as presumably they are for many top predators, site fidelity, if not full territoriality, seems to be the rule. The more uneven the distribution of resources, the less there is to be gained from site fidelity, and the more incentive there is to expand the area searched in the prospect of discovering rich rewards.

Tolerance of Neighboring Troops

Intolerance of neighbors is usually associated with territoriality; but there are other situations, such as the monopolization of temporary resource patches by howlers, that are mediated by intolerance that is not conditioned upon stable boundaries (Milton 1980). When ranked on their demonstrated tolerance vs. intolerance of neighboring groups, the five species again form an orderly series, but with one quirk. The *Saguinus* are fanatically territorial, *Cebus apella* normally tolerates other groups at close range, *C. albifrons* avoids close approaches to conspecific troops, and *Saimiri* is completely tolerant. The familiar ranking from *Saguinus* to *Saimiri* emerges once more except that the positions of the two *Cebus* species seem to be reversed. I shall now venture to rationalize the behaviors in terms of the differing life styles of the species.

Saguinus. The tamarins are extreme in their reliance on small common resources with a high degree of spatial and temporal continuity. Although dependence on resources with these properties has often been assumed to lead to territoriality, the connections can be clarified by examining the circumstances a little more closely.

An intriguing facet of the territorial behavior of *S. imperator* and *S. fuscicollis* groups is their evident reluctance to cross territorial boundaries, even in the absence of opposition. Instead of seizing opportunities, the troops usually announce their presence when approaching a border, and then wait, presumably to see whether their neighbors will respond. Even when no response is forthcoming, the animals nearly always retire to the interior of their territories.

Why don't they move in and help themselves to some of their neighbors' resources? It can plausibly be argued that they would gain no net benefit from doing so. Upon crossing the border they would enter unfamiliar terrain in which they had no prior knowledge of the locations of resource trees. Instead of going directly from one tree to the next, as they do within their own territories, they would have to hunt for them. Upon finding one they would face a good chance that it had been visited recently by the resident troop, and would consequently be empty of ripe fruit or nectar. The prospect of having to search at random for resources of uncertain payoff, as well as having to risk being discovered and attacked, is one that would appear desirable only under desperate circumstances.

As for the concentrated use of the central portion of their territories, at least three mutually compatible and reinforcing explanations are possible: to economize on travel, to minimize the risk of visiting trees that had been raided by interloping neighbors, and to remain within earshot of as much of the periphery as possible. Typical *Saguinus* territories are 400–800 m across. From the center of such an area, it is possible that an alert troop could hear a vocal challenge from any of its borders.

The territorial system of *Saguinus imperator* and *S. fuscicollis* serves the interests of both territory holder and neighbor alike. Cheating at territorial boundaries offers little or no advantage to either side, except perhaps in extreme cases of resource failure. Under most circumstances it is likely that a territory produces a substantial surplus of resources beyond the metabolic needs of the territory holder (Carpenter and MacMillen 1976). Normally, maximum foraging success will be attained within the home territory as the direct consequence of an intimate knowledge of the locations of resource trees, their rates of production of ripe fruit, and the schedule on which each tree had last been harvested. A high degree of predictability of resources, and a knowledge of where the resources can be obtained, are key ingredients in the self-interest of the territory holder; they create the perceived value of the territory, and they make it worth the price of defense.

Cebus apella. Now that we have gained some understanding of why the tamarins are so strongly territorial, we must ask why *Cebus apella* and the other species do not show the same tendencies. At a general level we suppose that the dichotomy of behaviors results from the fact that the tamarins subsist on diffuse, self-renewing resources while the remaining species subsist on more concentrated, nonrenewable resources. Nevertheless, there are differences of degree in the behaviors shown by the two *Cebus* species and *Saimiri* that require further analysis if we are to achieve a full understanding of their reactions to conspecific troops.

Cebus apella home ranges overlap broadly, especially during the dry season months when fruit resources are at their annual nadir. One might expect intraspecific competition and resource defense to be particularly pronounced at this time of year; instead we see what appears like a reciprocal granting of trespass licenses. Each troop expands its search radius to include the centers of its neighbors' home ranges. Intertroop encounters occur almost daily during this period. Although these encounters may occasion a certain amount of excitement among the animals, they normally do not lead to open antagonism, nor to other major perturbations in the routines of the troops, such as hasty abandonment of the encounter zone or retreat toward the center of the home range. One or both of the troops may simply move somewhat away from the other and continue its activity, or the two troops may even remain within sight of one another for times varying from a few minutes to an hour or more.

Intertroop aggression occurs almost exclusively near or in fruit trees. Two troops may occasionally feed amicably for an hour or more in the huge crown of a *Ficus perforata*, but when smaller trees are at issue, one group usually prevails through aggressive assertiveness. Interactions are especially intense at small but highly desirable resources such as *Scheelea* nut clusters. But even in contests over the possession of a *Scheelea*, the losing troop is not chased away, it is merely denied access until the winning troop is sated.

We see here a rather complex interaction in which space is shared, large resources (huge fig trees) are shared, but small resources (*Scheelea* palms) are not shared when discovered by two troops at once. Small resources are shared, however, in the sense that they are rather frequently discovered and exploited to exhaustion by troops that are well outside of the intensively used portions of their home ranges. How can the conditional tolerance or nontolerance of *C. apella* troops for each other be understood in terms of the self-interest of the individual troops?

Cebus apella is opportunistic in its use of big figs and other such bonanza resources. It may exploit them heavily for a few days, but in the aggregate, figs are a relatively minor component of the diet (see Chapter 5). The question we must answer is, What are the costs and benefits to a *C. apella* troop of defending a big fig? The costs would be measured as expended time and energy plus the risk of injury. The benefits are far less easily specified. Most of the competition for figs is preemptive competition. The crop is there for the taking, and first comers fare better than late comers. Figs are so popular with a wide array of birds and mammals that use of the resource is spread over many species, none of which harvests more than a minor fraction of the whole crop (Terborgh and Diamond 1970). No single species, no matter how determined, could possibly scare away all its competitors (Robertson et al. 1976). The attempt would present a ludicrous spectacle, like a naked man trying to fend off a swarm of mosquitoes. To be effective, the defense would have to be maintained on a 24-hour schedule, and for all that, the best that could be gained would be an extension of the period of availability. Since around-the-clock defense is out of the question, the only realistic alternative is to ignore other users and to feed as heavily as possible. It thus seems clear why *apella* tolerate each other at big figs: in relation to the size of the resource, the decrement that would be lost to another troop is too small to be worth fighting about, and the cost in lost foraging time would be as large as the benefit. The result is a laissez-faire attitude toward large fruit concentrations that works to the benefit of all.

But what about *Scheelea* palms, which are also an important dry season resource? These palms are quite regularly distributed at densities of 15–30 per ha over the entire study area (Kiltie 1980). It appears as if the production of nut clusters is nearly aseasonal, for we have records of their use by *C. apella* in every sample (January, April, June, August, October). *Scheelea* nut clusters thus appear to constitute a uniformly distributed, continuously available resource. One could easily suppose that dependence on them could lead to territorial defense rather than to home range overlap and sharing. The situation seems paradoxical. At issue are the losses that might be sustained through the incursions of alien troops. The magnitude of such losses could depend on whether ripe *Scheelea* nut clusters are harvested by chance or by design. If ripe clusters are exploited when found, and if unripe clusters are rejected and forgotten, then the process is purely one of chance discovery, and all troops are on a more or less equal footing, whether at home or in alien terrain. But if individual palms within the home range center are regularly monitored for developing nut clusters, and if these are harvested when the clusters reach an acceptable stage of ripeness, then discovery is not involved, at least not on the part of the resident troop. The knowledge gained through monitoring would clearly constitute an investment, because the monitoring would require time and energy, and the resultant knowledge would allow a more even and bountiful harvest than would accrue through chance discoveries. But by the same reasoning, monitoring would also confer to the resident troop a decisive advantage over an intruder, because it would be able to harvest nut clusters as soon as they ripened. If so, losses to intruders might be tolerably small. We are under the strong impression that *apella* troops do monitor resources, but without further investigation it is not possible to quantify the attendant costs and benefits. In their dry season wanderings, *apella* troops seem primarily motivated by the prospect of discovering large figs, and in this they may not constitute a sufficient threat to their neighbors to warrant violence. In any case, to both resident and alien troops, the considerations involved are complex and interesting, and merit further attention.

Cebus albifrons. By the logic employed above, *Cebus albifrons* troops should tolerate one another. Far more than *apella*, *albifrons* depends on concentrated resources, which we have just said cannot be effectively defended. What is the difference, and why should *albifrons* troops show such an aversion to one another?

Albifrons, unlike *apella*, uses its home range in a markedly patchy fashion. This results from its attraction to and lingering in areas of temporarily high resource density. These areas are of several kinds. At one extreme a resource patch can consist of a single huge fig (e.g., June sample). Less concentrated patches are formed by distinctive habitats with fruiting peaks that are offset from those of other habitats (e.g., the lake margin and riparian *Cecropia* zone, March sample). Or finally, a patch may consist simply of a closely spaced group of fruiting trees of various species (November sample). Once an *albifrons* troop has discovered such a patch, it tends to remain there until the fruit supply is largely depleted before moving to a new location, often many hundreds of meters away.

Given this pattern of resource exploitation, can the reaction of different troops to each other be understood? The costs of repelling another troop were considered above in the discussion of *Cebus apella*. I assume the costs are similar for *albifrons*. But what about the benefits? These relate directly to the advantage gained in possessing the resource. In the case of large figs, the advantage is small because the use is spread widely over many consumer species. But the advantage may not be so small as in the case of *apella*, because *albifrons* troops are larger and because figs assume a much greater importance in their overall food intake. For other types of resource patches the advantage will vary, depending upon the degree to which the resource is actually protected by driving away a rival group. The advantage would take the form of prolonging the period of exploitation before the resource was exhausted and a move became necessary. So long as the number of unexploited resource patches in the environment exceeded the number of *albifrons* troops, the troops would gain a mutual advantage in avoiding one another. Each troop could enjoy the full use of any patch it discovered, and the amount of searching for new patches, as well as the number of interpatch trips, would be minimized. As in the sharing of space and giant resources by *C. apella*, the observed behavior is one that appears to operate to the mutual benefit of both parties.

While mutual spacing may ordinarily be advantageous to *albifrons* troops, there must be occasions when it is not, such as in periods when the density of resource patches is less than the density of troops. These situations can be expected to arise most frequently in the dry season when the principal, if not the only, concentrated resource patches are fig trees. Nearly every time we have had occasion to observe the activity around a large fruiting fig in the dry season, it has been visited by two *albifrons* troops. Typically one (the dominant troop?) has fed in the early morning for an hour or two and then departed for its daily foraging run. The second troop has then appeared in the middle of the day or early afternoon when it has enjoyed an equally long feeding bout. Thus, neighboring *albifrons* troops do occasionally share resources; but the sharing is always sequential and never amicable, as it usually is in *apella*.

Cebus albifrons defends space, but, like howlers, it defends only the resource patch it is currently using (Milton 1980). Why not a territory? One frequent rejoinder is that the area used is too big to defend, in this case more than 2 km across and several times the distance that a loud vocalization carries (cf. Mitani and Rodman 1979). But the same argument could be applied just as well to the larger *Saguinus* territories (up to 70 ha). Even though the boundaries may be visited and advertised only occasionally, a rigid territoriality is maintained, as we speculated, by the mutual benefits realized by both neighbors. In the case of *Cebus albifrons* there may be no such mutual benefit; if so, the size of the area involved is irrelevant. What is important in the case of *albifrons* is that it uses large resource concentrations that often appear at unpredictable times and places. For a species depending on such resources there is no advantage in living within a rigidly bounded space, for the next resource patch to appear in the region may be outside the bounds. Nevertheless, *albifrons* troops generally remain within a hazily defined area. Why?

In the absence of rigid boundaries, two inhibitions remain to create a more or less well-defined home range. One is the fear of encountering hostile neighbors and the other is the handicap almost any intelligent animal must concede when searching for food in unfamiliar terrain. These constraints act to define a center of gravity for each *albifrons* troop, and assure a more or less even spacing of their centers of gravity, even though their outer limits may be ill defined. In its "sharing" of space, *albifrons* seems to go even farther than *apella*, and thus in one sense is less "territorial" in spite of the mutual avoidance behavior. At least this is what is implied by the indirect estimates of home range overlap that were calculated in Table 7.1. How one ranks the two *Cebus* species on a scale of territoriality depends on whether greater weight is given to the intensity of antagonism and repulsion that accompany encounters, or to the realized amount of spatial overlap and its implied sharing of resources.

Saimiri. We come now, finally, to the squirrel monkey and its behavior in troop encounters. As mentioned before, these encounters are always peaceful, at least in outward appearance. But *Saimiri* depends on concentrated resources even more than *albifrons*, and if *albifrons* troops avoid one another, why don't those of *Saimiri*?

The question can perhaps be answered at two levels. At a proximate level, it may be because there is no individual or group of individuals in a *Saimiri* troop that takes responsibility for the whole. The groups have no obvious internal hierarchy or individuals with recognizable leadership roles. We have looked for evidence of the kind of organization that is so apparent in *Cebus* but failed to find it. Instead of retaining a definite composition for long periods, it is our impression that *Saimiri* troops frequently exchange members. However, we have no way of knowing this with any certainty in the absence of marked individuals.

Another peculiarity of *Saimiri* is that group defense behavior is lacking. Both sentinel behavior and overt defense (mobbing, close approach threats) are well developed in *Saguinus* and *Cebus*, but the species in these genera have troop structures. Among *Saimiri* no such unity is apparent, and every individual seems to fend for itself.

These apparent behavioral deficiencies are a moot issue most of the time. As pointed out in Chapter 8, *Saimiri* live much of their lives in the company of *Cebus* in a relationship that can be called commensal or mildly parasitic. Within the mixed association, *Cebus* lead, *Cebus* sound most of the alarms, *Cebus* defend against predators, and *Cebus* battle over rights to particular feeding places. The *Saimiri* are merely hangers-on, receiving the benefits of *Cebus* organization and aggressiveness without contributing themselves. One could almost imagine that they are behaviorally degenerate in certain respects, as true parasites often are.

While a proximal explanation for the lack of intertroop aggressiveness in *Saimiri* may derive from the apparent absence of organized leadership in the groups, the ultimate explanation may have more to do with the species' small body size and the fact that it eats common fruits that are consumed by many other monkeys. Large ephemeral resources that are used by many species are probably not worth defending, as we noted in the discussion under *Cebus apella*, but this would especially be true for a species that is low in the interspecific dominance hierarchy. It is hard to see how *Saimiri* could gain any advantage by fighting each other over a tree that was also being exploited by *Alouatta*, *Ateles*, two species of *Cebus*, etc. The situation of *Cebus albifrons*, however, is not the same, as they are capable of dominating all the other species except for *Ateles*. *Saguinus* also gain an advantage in fighting over their resources, because, as I showed in Chapter 5, those resources are not much used by larger monkeys.

Now, to sum up the arguments. Each species displays a distinct mode of spatial utilization, and along with it a different characteristic reaction to conspecific troops. The several ways of using space seem to be adaptations to exploiting fruit resources that appear in the environment with varying degrees of temporal and spatial patchiness. In turn, the types of interactions that occur when troops meet can be understood in terms of the costs and benefits of defending resources that have different scales of patchiness.

The resources used by *Saguinus* have a high degree of predictability and continuity in both time and space. Territorial recognition and defense is mutually advantageous to both resident and neighbor alike, because under normal circumstances cheating (incursions into a neighbor's space) is a low-benefit, high-cost proposition.

Cebus apella uses space flexibly, expanding out of a consistent core home range to double or triple the area covered during periods of fruit scarcity. Its diet consists of a mixture of nonpatchy (palms) and patchy resources (figs and other large fruit trees). Intertroop encounters are normally nonaggressive, and adaptations for controlling space, such as loud calls, appear to be poorly developed. Each *C. apella* troop shares its home range with several others in a practice that maximizes their mutual benefit in the harvest of large resources, such as figs. More evenly distributed resources are harvested mainly within the intensively covered central portion of the home range.

Cebus albifrons, to a much greater degree than *apella*, depends on spatially restricted concentrations of fruit. Defense of such patches can lead to prolonged periods of utilization and less frequent moves between patches. By means of loud calls delivered at a distance, rarely by face-to-face intimidation, *albifrons* troops fend other troops away from current feeding areas. This is a form of territoriality in which hoards of nonrenewable resources, not space as such, are defended. It is in some ways analogous to the situation that is prevalent among nectar-feeding birds in which a patch of flowers is defended against conspecifics and often against other species as well (Gill and Wolf 1975; Carpenter and MacMillen 1976). The difference is that the daily yield from a patch of flowers can be enhanced by defense, but the period of use of the patch as a whole cannot be. With *albifrons* it is the period of use that is at contention, not so much the immediate rate of harvest.

Finally there is the case of *Saimiri*, which is even more a temporary patch specialist than *Cebus albifrons*, but which is nonaggressive and does not defend its resources. The reason for this seems to stem from the animal's small size and the consequent futility of fighting conspecifics when much larger species are responsible for the main inroads on its food supply.

Group Size in Primates

Over the past decade primatologists have shown great interest in trying to understand the selective forces that shape the organization of primate societies (Aldrich-Blake 1970; Clutton-Brock 1974; Clutton-Brock and Harvey 1977a, 1977b; Crook and Gartlan 1966; Crook 1972; Eisenberg et al. 1972; Hladik 1975; Jolly 1972; Struhsaker 1969; Sussman 1977). Although many important patterns have been brought to light by these studies, I wish to draw attention to four correlations that have particular bearing on the problem of group size:

(1) home range area is strongly proportional to group size in interspecific comparisons (Clutton-Brock and Harvey 1978); (2) nocturnal primates are nearly always solitary (Jolly 1972; Charles-Dominique 1977; Pollock 1979); (3) primates that occupy open habitats live in large groups (Crook and Gartlan 1966; Clutton-Brock and Harvey 1977a); (4) among arboreal species, ones that subsist on clumped resources tend to have larger group sizes and larger home ranges than ones that feed on more uniform resources (Clutton-Brock 1974, 1975; Hladik 1975; Sussman 1977; Waser 1977). I shall have suggestions to make about the causalities behind all these generalizations, but before that some preliminary remarks are in order.

I shall assume that pronounced differences between species in social traits are due to natural selection just as much as differences in morphology. Even in questions of morphological adaptation it is often unclear what the selective forces are—why an animal is a given size, for example (Clutton-Brock and Harvey 1977a). Usually one assumes that the relevant selective forces are external to the species in question, i.e., that they derive from interactions with the physical environment, food supply, or other species. But in considering social adaptation, even this elementary assumption is not assured. Sociality, in itself, creates conditions within which an entirely new array of selective forces can operate: intrasexual competition for mates, the struggle for rank in dominance hierarchies, parent-offspring conflicts, benefits derived through inclusive fitness, etc. (Alexander 1974; Wilson 1975). Each category of competitive or cooperative interaction between individuals provides a whole spectrum of adaptive choices, and these adaptive choices may be influenced largely by conditions that exist within the population, as distinct from external forces.

But it is clear that social systems do not evolve solely in response to conditions that already exist within a population. If this were so, it would be very hard to understand the undeniable correlations that do exist, for example, between the social systems of antelopes and their habitats (Jarmon 1974; Estes 1974), between those of birds and their diets (Crook 1965; Horn 1968; Orians 1969a), or between those of primates and habitat or trophic status (Eisenberg et al. 1972; Aldrich-Blake 1970; Crook and Gartlan 1966; Denham 1971; Kummer 1971; Struhsaker 1969; Hladik 1975; Jolly 1972; Clutton-Brock and Harvey 1977a, 1977b). It would also be hard to understand the diversification of social systems within single monophyletic lineages such as the Cebids or Bovids. We are thus faced with the inescapable inference that the external environment can and does selectively modify and transform social systems. It is not necessary, however, to assume that all social traits are genetically fixed, or that the behavior of individuals may not be modified directly by interactions with the environment. The questions remain as to what features of the environment act to modify social systems, and what aspects of social systems are most immediately exposed to modification by the external environment.

Group Size and Ecology

I shall begin to answer these questions by considering first the problem of group size. Within the species studied, group size is closely associated with several ecological parameters: home range area, home range overlap and its inverse, site fidelity, degree of dependence on figs, and mean size of fruit trees used (Table 10.1). It is highly unlikely that the coincident rankings of all these variables is merely due to chance. But what are the underlying relationships?

It has often been observed that animals which live in large groups have larger home ranges than related species which live in small groups (e.g., McNab 1963; Jarmon 1974; Clutton-Brock and Harvey 1977a). In graphical presentations of data, some measure of group size has often been scaled on the horizontal axis, and home range area on the vertical axis. By plotting the data in this way it is implied that group size is the independent variable. Indeed, this seems like the reasonable thing to do, because a large group of animals obviously requires more resources, and hence more space, than a small group. There can be little doubt that this is the correct way to view *intra*specific comparisons (e.g., Waser 1977). But it is not clear that group size is the independent variable in *inter*specific comparisons.

Suppose we turn the question around and imagine that home range area is the independent variable. Now a different picture begins to emerge. If we assume as the starting point that an animal's diet is its most fundamental adaptation, then it is evident that a species which depends on scattered,

Table 10.1
Rank order of group size and other socioecological parameters of the primates studied (low to high).

Species	*Mean group size*	*Home range size*	*Home range overlap*	*Site fidelity*	*Dependence on figs*	*Mean fruit tree size*
Saguinus spp.	1	1	1	4	1	1
Cebus apella	2	2	2	3	2	2
C. albifrons	3	3	3	2	3	3
Saimiri sciureus	4	4	4	1	4	4

concentrated resources must have a large home range. Since clumped resources by their very nature are highly productive, it is possible for many individuals to exploit them at once without experiencing undue interference (Brown 1964). Such resources can be considered permissive of large group sizes. They provide an extrinsic motive for the peaceful aggregation of large numbers of conspecifics, but in themselves do not provide the conditions for permanent sociality, i.e., cohesive groupings of individuals. Some further impetus must therefore be involved in the evolution of sociality.

Advantages of Living in Groups

So far we have derived the conditions for passive aggregation at concentrated food sources. What further incentives to sociality could make it advantageous for individuals to remain together after feeding, or after the original food source was exhausted? Several such incentives have been proposed; but broadly speaking, they fall into two categories: protection against predation, and social facilitation of foraging (Alexander 1974).

Evidence for the influence of predation on the adaptedness of cohesive groupings can be deduced from comparing the behavior of diurnal and nocturnal animals. Nearly all nocturnal animals, apart from bats, are solitary. Many diurnal birds and mammals are gregarious. The contrast can be understood if we consider the sensory modalities used by nocturnal and diurnal predators. Most nocturnal predators hunt by sound (Payne 1971; Konishi 1973; Curio 1976). This is the appropriate modality under conditions of poor to zero visibility and low background noise (lack of convectional winds). Attack is either by stealth (owls) or by ambush (felids). In either circumstance, the predator is inaudible and invisible to the prey until the last instant, and therefore no early warning system is possible. A prey animal avoids attack by being silent. Noise levels resulting from foraging or traveling will inevitably increase with group size (Kiltie 1981). Groups are thereby exposed to increased risk, and solitary foraging is therefore the rule. By day, predators hunt mainly by sight and capture their prey after violent pursuit. A prey animal can usually escape if it spots the predator soon enough. Increased group size, by augmenting the number of attentive eyes, can result in enhanced vigilance at lower per capita cost in wasted time to foraging individuals (Pulliam 1973; Powell 1974; Siegfried and Underhill 1975; Bertram 1978; Caraco 1979). This benefit plus the reduced per capita probability of being killed in an attack (safety in numbers) combine to enhance the value of groups to their members among diurnal species.

In addition to probable benefits in reducing predation, sociality carries some potential advantages in facilitating the task of food finding. Inexperienced individuals could gain from following experienced ones, just as *Saimiri* appear to profit from following *Cebus*. In this case the benefit to the experienced individuals is negative, in that food discoveries would have to be shared more widely. Hangers-on could pay their way, however, either by contributing to a selfish herd (Hamilton 1971; Vine 1971) or by increasing the rate of discovery of new food resources. Among the primates we studied, however, it is doubtful that important resource patches are often encountered by chance. Either they are well known to experienced individuals, as in the case of seasonally fruiting habitats (e.g., fig swamps), or they can be detected at a distance from the shrieks of parrots (large figs). This leaves us with protection from predation as the most compelling initial impetus to sociality.

Disadvantages of Groups and the Fixing of an Optimal Group Size

Mutual interest in lowering the risk of predation can lead to cohesive aggregations of individuals; but as an explanation of sociality, it does not offer any means for understanding the varied sizes of social groups. This is because, other things being equal, individuals are afforded better protection in large than in small groups. There must be other, quite different factors that act to limit the numbers in groups and thus to define the optima that characterize the typical group sizes of different species. On this point there is a broadening consensus that group size is often limited by intragroup competition over access to feeding sites (Krebs 1974; Pulliam et al. 1974; Dittus 1977; Wrangham 1977; Caraco 1979; Struhsaker and Leland 1979). But if this is so, why should intragroup interference build to intolerable levels at such different group sizes in different species? This can be explained if we assume that each species feeds on resources having a certain characteristic level of concentration, or in the jargon of ecology, patch size. Aggression will build up quickly with group size in species that feed on small resources, and only at much greater numbers in species that typically feed on large resources. I propose that group size is set by an optimal trade-off between increased predator protection and decreased foraging opportunity, exacerbated by greater travel cost (Fig. 10.1). The model explicitly holds that social groups are supraoptimal for maximal foraging success. Specifically, it is expected that the decrement in survival potential that an animal experiences in switching from a large to a smaller group is, on the average, offset by an equal or greater increment in survival or reproductive potential by the better feeding opportunities offered in the smaller group.

This model is similar in some respects to one proposed by Pulliam (1976) and later, in more elaborate form, by Caraco (1979). In their theory, emphasis is placed on additional feeding time gained through increasing group size as a result of the decreased need for vigilance by individual foragers. The ultimate

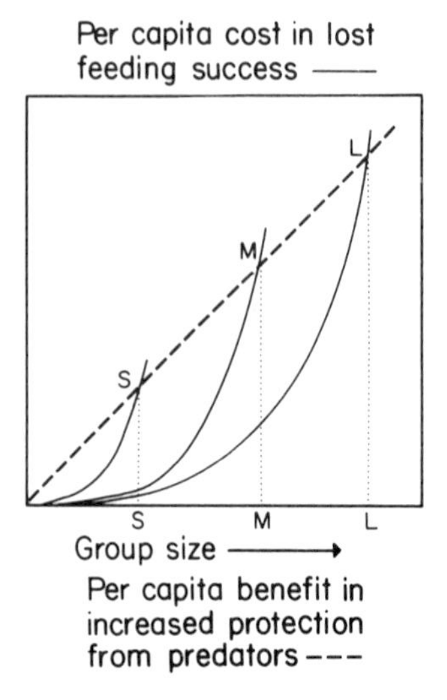

Figure 10.1 Model for the regulation of group size in primates through the interactions of resource size, individual foraging efficiency, and the benefit of groups in predator detection and avoidance. The left ordinate represents the cost of joining a group of size N in terms of lost feeding opportunity, ultimately expressed as reduced survival and/or reproductive success. The right ordinate represents the enhanced safety conferred by joining a group of size N, ultimately expressed as improved survival. The feeding cost incurred at any group size depends on the characteristic resource size of the species. Cases are illustrated for small, medium, and large resources, leading respectively to small, medium, and large optimal groups.

impetus for aggregation, however, is fear of predation. Increasing strife over food resources imposes the upper limit to group size. The model presented here contains the additional feature that *inter*specific variation in group size is explained as an adjustment to the characteristic patch size in which the resources of any species are distributed. The model is also reminiscent of ones presented by Wittenberger (1980) except that it stresses explicit proximal causes (predator protection and intragroup competition for food) rather than more vague ultimate causes (increased survival vs. decreased reproduction with increasing group size).

Relationships between Group Size and Ecological Parameters

The line of reasoning that led to the formulation of the optimal group size model was derived from consideration of the coincident trends in Table 10.1. It did not seem reasonable that the parallelism in the variables was due to chance. The problem then was to decide which of the variables, if any, was the independent variable. If one chose group size, as had previously been done, then group size could not be explained by any of the other variables. But, if one chose instead the size of the average resource patch, as represented by mean fruit tree size and less directly by the degree of dependence on figs, then a coherent picture could be pieced together. Large resource concentrations are necessarily rare, and therefore any animal that harvests them with regularity must have a large home range. Home range size will thus scale in some increasing fashion with increasing spatio-temporal patchiness of resources. An increasing tendency to wander in search of rare resource concentrations will imply decreased site fidelity and increased home range overlap, for reasons given in the first part of this chapter. Finally, group size can be accounted for as a passive response to the size of the resource or number of available feeding sites (Leighton and Leighton 1982).

A Thought Experiment

The relationship between foraging ecology and group size in the species we studied can be further illuminated by conducting a thought experiment. I shall do this by isolating the variable of group size from its coadaptive correlates in contemplating a role reversal of *Saimiri* and *Saguinus*. I shall assume that the diets and population densities of both species remain constant, but that each takes on the social characteristics of the other.

First, let us consider the difficulties that *Saimiri* would encounter if it lived like *Saguinus*. Instead of traveling far and wide in throngs of 30 to 40, the individuals would be grouped in sedentary, territorial family units of five. At a population density of 60 per km^2, each family would have about eight hectares at its disposal. So long as fruit were plentiful, this would probably be an adequate amount of living space. But the animals would be in an untenable situation as soon as the fruit crop dwindled. Figs, their principal dry season energy source, are widely scattered. A generous estimate of the density of ripe crops during the May–July critical period would be one or two per km^2 (see Fig. 6.1). Most territories would be bare of fruit for months on end. Not a happy picture.

Now, let us turn the tables and imagine what would happen if *Saguinus* were to live like *Saimiri* in groups of 35. If a mixed group of 7 *Saguinus* has to spend 16% of its time feeding in the 8–10 trees it visits each day, plus another 22% traveling between them, there would doubtfully be enough time in a day to visit the 40–50 trees that would be required to sustain so large a group. There would be other adversities as well. Instead of spending nearly half of each day huddling in dense vine tangles where they are beyond the reach of aerial predators, they would now have to be on the move almost continuously, thereby proportionately increasing their exposure to predation.

Moreover, if the groups were to wander seminomadically over many km^2 of forest in the fashion of *Saimiri*, their foraging efficiency would be compromised on two counts. First, they would have to spend more time and energy searching for producing trees in unfamiliar or seldom-visited terrain. And second, they would have to traverse large areas of habitat that are not occupied by *Saguinus*, presumably because they are in some way unsuitable. In short, neither species could assume the social system of the other without acquiring unacceptable ecological handicaps.

The thought experiment thus provides a clear demonstration of the crucial roles of group size and home range area as determinants of foraging success. In *Saguinus* the groups are small because any increase in group size entails losses in foraging efficiency, and imposes increased travel requirements and concomitantly greater risks of predation. In *Saimiri*, seminomadic behavior is essential to harvesting figs. The large groups of *Saimiri* do not contribute in any evident way to foraging success. Rather, they appear primarily to serve as an antipredator adaptation by reducing the per capita risk of attack. Other means of achieving the same end are closed to *Saimiri*. It is too small to defend itself against a formidable array of large hawks and eagles, and in its incessant traveling and foraging, it is too busy to spend much of the day in hiding as *Saguinus* does. When predators are evaded through crypsis as in *Saguinus*, selection favors small groups, but when the evasion is through safety in numbers as in *Saimiri*, individuals maximize their security by joining the largest groups that are commensurate with meeting adequate feeding requirements (e.g., Kiltie 1981).

Tests of the Theory: Exceptions That Prove the Rule

The model suggests two straightforward types of predictions, each representing an extreme situation: (1) species that are not threatened with predation should forage solitarily, or at most in small (female-offspring) units; (2) species that are not limited in their feeding by the size of resource patches should aggregate in large numbers in the presence of predators. I shall now examine both of these propositions.

It appears that there are a number of arboreal primates that are too large to be taken by aerial predators, at least as adults. In the New World these include the species in *Ateles*, *Brachateles*, and *Lagothrix*, all of which weigh 8 kg or more as adults. Even though the members of all three genera feed on clumped resources (large fruit trees), none of them has a highly cohesive group structure (pers. obs., and fide K. Milton for *Brachateles*). They live in fission-fusion societies. Individuals aggregate, sometimes in considerable numbers, at large resource concentrations and disperse when feeding in scattered small trees (Klein and Klein 1975, 1977). Howlers (*Alouatta* spp.) appear to represent an exception in that they are large (ca. 8 kg) but live in closely integrated groups. However, howlers are far less swift and agile than either *Ateles* or *Lagothrix* and are known to be preyed upon by harpy eagles (Rettig 1978). The finding of fission-fusion societies exclusively among the largest, best protected species supports the case for threat of predation as the primary social cement.

Further examples of fission-fusion societies can be found in the Old World primates, and again the species are of exceptional size. They are the chimpanzee (*Pan*: Wrangham 1977) and the orangutan (*Pongo*: Rodman 1977). Among chimpanzees the advantage of small groups during feeding was noted by Wrangham (1977) who found that the feeding time per individual decreased as a function of group size. Both chimps and orangs are far too large for any extant aerial predator.

Although gorillas are even larger, they spend most of their time on the ground where young individuals are potentially subject to being ambushed by big cats. It is not known whether the siamang and the gibbons are taken by aerial predators. Even if they are, their relatively modest space requirements and territorial behavior suggest that they do not depend on highly clumped resources (Chivers 1973, 1977; Curtin and Chivers 1978).

Among the Old World monkeys there is only one species that appears to have a true fission-fusion society, and that is the hamadryas baboon (*Papio hamadryas*: Kummer 1968). Subgroups disperse for feeding and aggregate for the night at sleeping cliffs, presumably to escape nocturnal predators (felids) in a virtually treeless landscape. Although the groups of several additional species are reported to fragment at times during feeding, the troops progress as units and have a definite long-term membership (e.g., *Mandrillus leucophaeus*, *Nasalis larvatus* (fide Clutton-Brock and Harvey 1977a); *Papio cynocephalus* (Altmann 1974); *Presbytis obscurus* (Curtin and Chivers 1978); *Macaca fascicularis* (Aldrich-Blake 1980); *Cercocebus albigena* (Waser 1977). In general, the fact that small group sizes (less than 10) are extremely rare in Old World monkeys suggests that all of them are exposed to some threat of predation. More study of many species in their undisturbed natural environments will be needed to reject or support this contention.

Among frugivorous and folivorous arboreal primates, the number of feeding sites at a resource is necessarily limited by the size of the largest trees. Maximum group size in such primates is accordingly constrained, seldom exceeding 40 individuals (Jolly 1972). However, among terrestrial species, especially those frequenting open country, there is no such constraint, and group sizes larger than 40 are fairly routine. In some species there is evidence of facultative increases in group size with the arboreal-terrestrial transition, as was noted by Hladik (1975) in grey langurs when they became more terrestrial during the dry season. Truly enormous aggregations of animals, such as those seen in ungulates or waterfowl, are virtually restricted to open country. There, the threat of predation is high due to the unhampered visibility of the prey, and for herbivores the number of feeding sites in a grassy sward or marsh is essentially unlimited. Competition for feeding sites is therefore minimal, and the tendency to seek safety in numbers takes over to promote aggregations of remarkable size.

We can thus conclude that both tests of the model are generally upheld: diurnal animals that are free of the threat of predation tend to forage alone or in small family units, and animals that live in open country and feed on resources that are not spatially restricted tend to form large aggregations. In between the two extremes the theory offers a great latitude for understanding the adaptive significance of intermediate-sized groups, as the relative strengths of predation pressure and intragroup aggression vary from one species to another.

Group Size and Primate Social Systems

I wish to conclude this chapter by addressing one last issue: whether there is any necessary connection between group size and social system. The evidence from primates strongly implies that there is (Fig. 10.2). By fixing group size, the number of possible social systems is narrowed to a very limited number of alternatives. Monogamy occurs in species whose characteristic group sizes range from 2 to 5. The smallest single male polygynous groups occur in such species as *Alouatta seniculus*, *Colobus guereza*, and *Cercopithecus neglectus* with group sizes of 6 to 12, while the largest such groups contain between 15 and 25 individuals (*Presbytis* spp., *Cercopithecus* spp., *Erythrocebus*). Stable groups containing more than about 25 individuals virtually always have a dominance-stratified multimale structure (e.g., *Saimiri*, *Miopithecus*, *Colobus badius*, *Macaca* spp., *Papio* spp.). It can thus be asserted that, to a first approximation, group size does define the social system, except within the range of 10 to 25 individuals where both single and multimale systems are found.

In an evolutionary sense the single male (harem) and multimale polygynous systems are undoubtedly very close. While monogamy and polygyny, for example, are never found in the same genus, both single and multimale groups are found in several genera (e.g., *Colobus guereza* vs. *C. badius*, *Alouatta seniculus* vs. *A. palliata*, *Presbytis senex* vs. *P. entellus*). And in at least some species of langurs, both types of social organization occur within continuous populations (Oppenheimer 1977; Curtin 1980).

Let us consider the trade-offs that may be involved in selecting for the two

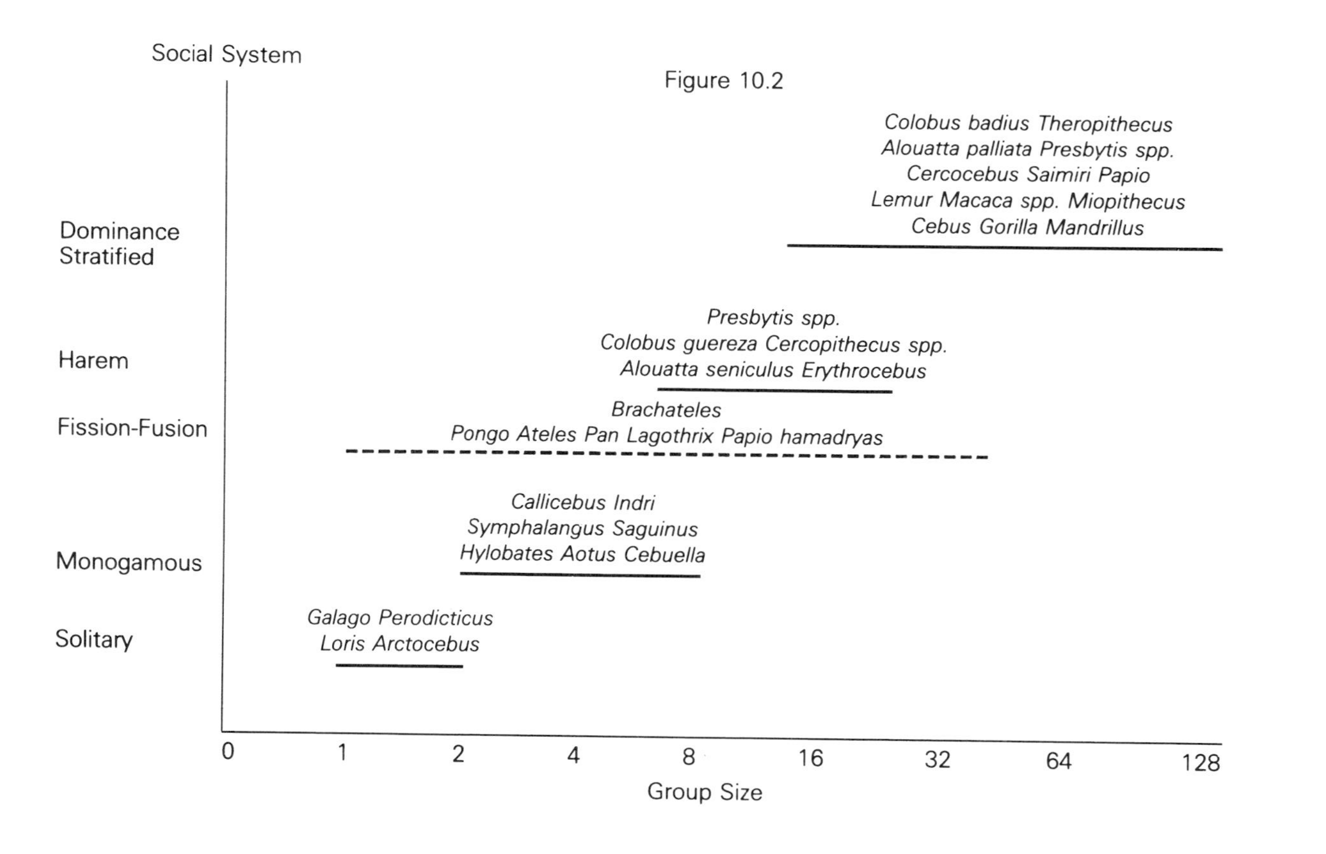
Social System
Dominance
Stratified
Harem
Fission-Fusion
Monogamous
Solitary
Colobus badius Theropithecus
Alouatta palliata Presbytis spp.
Cercocebus Saimiri Papio
Lemur Macaca spp. Miopithecus
Cebus Gorilla Mandrillus
Presbytis spp.
Colobus guereza Cercopithecus spp.
Alouatta seniculus Erythrocebus
Brachateles
Pongo Ateles Pan Lagothrix Papio hamadryas
Callicebus Indri
Symphalangus Saguinus
Hylobates Aotus Cebuella
Galago Perodicticus
Loris Arctocebus
0
1
2
4
8
16
32
64
128
Group Size

Figure 10.2

systems. Where either reduced predation pressure or uniform resources favored small groups, single males might succeed in monopolizing cohorts of females. An increase in predation pressure would heighten the vulnerability of unassociated cohorts of males, and thereby increase their resistance to being expelled by the dominant male. This is one impetus that could lead to multimale groups. Secondary males may be tolerated if they form a coalition with the dominant and help defend the group against take-over by extra-group males, thereby improving the prospects of all members of the coalition for long tenure in the group. In some species this appears to be a conditional mechanism that depends on the number of extra-group males in the population (Rudran 1979). Use of clumped resources by reducing internal competition for food is the mechanism that best seems to account for the large home ranges and correspondingly large group sizes of several species relative to sympatric congeners (e.g., *Colobus badius* vs. *C. guereza*; *Presbytis entellus* vs. *P. senex*; *Lemur catta* vs. *L. fulvus*). But how do we explain the fact that *Cebus apella* has multimale troops containing only 10–12 individuals, while the blue monkey (*Cercopithecus mitis*), for example, has single male troops with as many as 27 members (Rudran 1978)? It would seem reasonable to suppose that a single male would strive to monopolize a harem except when the associated costs were prohibitive, or when there were other mitigating circumstances in effect, such as cooperative male defense against predators or against other troops in contests over resources or access to females.

Cooperative male defense is well known in baboons (Altmann and Altmann 1970), and can be expected to occur in other species with multimale troops (Wrangham 1980). We have seen suggestions of it in *Cebus*, though our observations should not be regarded as definitive. Where both terrestrial and aerial predators are a threat, as in ground-dwelling and open-country species, multimale troops are the rule (*Papio*, *Theropithecus*, *Macaca*, *Lemur catta*). There is apparently only one species that does not conform to this generalization. It is *Erythrocebus* which has exceptionally large single male troops, but which hides in tall grass and flees at high speed when startled. The dominant male acts as sentinel, but does not defend the group (Hall 1965).

If one assumes that group size is the basic social parameter that is optimized by a species' interaction with its environment, the type of social system it has will follow more or less automatically. Where optimal group size does not predetermine the social system, as in the range of 10–25 individuals, the social system, whether single or multimale, will depend on alternatives available to the males, as discussed above. Since males can be expected to respond to variation in group size primarily as it affects their perceived opportunity to breed, the ultimate regulation of group size in primates probably depends on the degree of cohesiveness among females, as Wrangham (1980) has suggested.

While it can be claimed that the optimal group size model is strongly supported by the data examined, it would be procedurally incorrect to assert that it had been fully tested. The results obtained in the present study served to generate the model, and therefore cannot be used to test it. Further research must be done before the idea can be confirmed or rejected.

Any of several kinds of tests would be appropriate. Following the lead taken here, one might document resource size characteristics of another set of related species that differed in group size. A more direct, experimental approach would be relatively easy in intraspecific comparisons. One could, for example, study how group size varied with the intensity of predation pressure. (Pulliam et al. [1980] have done such a study with juncos using a trained hawk, and found that group size increased as expected.) Under decreased threat of predation it would be expected that group structure would relax and that individuals would forage much more often on their own. A number of predator-free islands onto which primates have been introduced could provide the test cases. The possibilities for further tests are legion.

Summary

The use of resources and space by the five species is reviewed in the opening section of the chapter. It is then noted that the reactions shown by groups toward neighboring conspecific groups are remarkably different in every case. The reactions range from complete tolerance (*Saimiri*), to conditional antagonism (*Cebus apella*), to strong mutual avoidance (*Cebus albifrons*), to vigorous territorial defense, including fighting (*Saguinus*). While the strengths of the interactions are easily ranked, the behavior of the species is not readily understood in terms of simple measures of territoriality or the use of space. The pattern can be understood, or at least rationalized, from a knowledge of what resources are most important to each species and how these resources are distributed in space and time, and from a consideration of the probable costs and benefits of defending key resources against conspecifics.

The potential interrelationships between group size, home range size and resource size are taken up next. I argue that a species' diet is a fundamental adaptation, and that the process of efficiently harvesting any given diet carries strong implications for the use of space. At one extreme, dependence on highly concentrated and widely scattered nonrenewable resources calls for seminomadic habits and a large, flexible home range, while at the other extreme, dependence on common small renewable resources leads to regular spatial utilization patterns and territoriality. The size of the home range an animal occupies is thus seen as being closely coupled to the characteristic temporal and spatial distribution pattern of its principal resources. Although group size often scales in parallel with home range area in interspecific comparisons, and does so in the present set of species, there seems to be no direct connection between the two variables. Instead, the large scattered resources that are typically exploited by species with large home ranges provide permissive conditions for group feeding. By this reasoning, use of concentrated resources is regarded as a necessary but not sufficient condition for the evolution of social groups.

The case is then developed that predation pressure provides the strongest and most universal impetus for sociality. Powerful evidence for this is found in the contrasting means of evading nocturnal and diurnal predators. Nocturnal predators use stealth and ambush to attack targets detected by sound, while diurnal predators use sight to detect targets that are attacked by violent pursuit. In the first case, being in a group is a liability because silence is the best refuge at night and groups inevitably make more noise than solitary individuals. In the second case, being in a group is an asset because by day a group offers safety in numbers and is better able to detect a predator before it attacks.

Aggregation as an antipredator tactic and the use of patchy resources provide the ingredients for a model of optimization of group size. Habitual exploitation of large food sources that provide many feeding sites permits the formation of large groups that are stabilized by the members' common interest in minimizing the risks of predation. Small food resources provide few feeding sites, and therefore animals that exploit such resources cannot join in large aggregations without experiencing unacceptable levels of intragroup competition. Such animals are thereby constrained to live in small groups. The typical group size of a species is thus seen as an optimal trade-off between an increasing benefit (lower per capita threat of predation) and an increasing cost (intensified intragroup competition for feeding sites), with resource size playing a crucial but passive role in imposing an upper limit on group size. Implicit in the model is the expectation that under the threat of predation, groups always tend to be too large for the simple maximization of feeding efficiency.

In the concluding section, the notion of an optimal group size is applied to the problem of understanding the adaptive basis of the varied social systems of primates, and by implication, other social vertebrates. It is assumed that the main selective influence that the external environment exerts on the structure of social groups is to regulate the numbers of individuals that live together. Much of the observed variation in primate social systems can then be explained as an automatic consequence of group size, for once group size is fixed, the number of alternative social systems that a species can adopt is extremely limited. Which among these alternatives evolution may favor in any particular case remains a difficult problem, as the choices must depend both on interactions with the external environment and on the social forces that operate within the species itself.

Epilogue

The seemingly limitless forest that stretches from the base of the Andes to the Atlantic Ocean is being encroached upon today as never before. Modern man is penetrating into the remotest corners of the Amazon basin in search of timber, minerals, and agricultural land, and vast development projects are already underway. The most valuable timber species have all but disappeared from the vicinity of roads and navigable rivers, and several species of animals have been driven to the brink of extinction. Animal populations generally have suffered drastic declines due to laissez-faire exploitation for meat, hides, and the live animal trade. Nearly everywhere "development" begins with clearing of the forest. Very little effort has been given to learning how to use the forest without completely eliminating it, and with it the immense diversity of life it contains.

From what has been learned in the present work one can begin to address the question of how the forest could be used without destroying its potential for supporting wildlife. Even selective cutting can do great damage if the species removed do not replace themselves, or if they provide an indispensable link in the annual food cycles of certain animals. In the forest at Cocha Cashu a very limited number of plant species sustains a major part of the animal biomass through the worst period of the dry season (Table 11.1). The plants are of various types—trees, vines, and palms, and their products are equally diverse—fruit, nectar, flowers, and nuts. Some (*Combretum*, *Erythrina*) are most common in the flood-disturbed zone along the river, while others are apparently confined to mature high-ground forest (*Celtis, Ficus* spp., *Quararibea*). The palms are common in nearly every habitat.

Species on the short list in Table 11.1 accounted for over 90% of the early dry season feeding time of all five primate species in the study, and other observations (e.g., Janson et al. 1981) suggest that they are equally important to many other animal species as well. What is most remarkable, though, is that this list of critical resources includes less than 1% of the plant species that have been identified at Cocha Cashu. A minuscule fraction of the botanical diversity thus plays a wholly disproportionate role in the food chain of the forest. One must wonder what would happen if some or all of these species were removed from the system. There just do not seem to be other resources available to take their places, so one can infer that the carrying capacity of the environment would be severely reduced for the animals listed in the table. The bright side of the picture, however, is that a knowledge of the roles these plants play in the ecosystem could open the door to effective habitat management in the future.

Table 11.1
Critical dry season resources in the Manu flora.

Plant species	Resource	Period available	Eaten by
Palms			
Astrocaryum sp.	Seed	May, June	*Cebus* spp., peccaries, squirrels, other rodents, macaws
Scheelea sp.	Mesocarp, seed	Throughout	*Cebus* spp., *Saimiri* (mesocarp); squirrels (seed)
Iriartea ventricosa	Exocarp and seed	May–July	*Callicebus moloch, Cebus apella, Ateles* (exocarp); peccaries (seed)
Vines			
Celtis iguanea	Fruit	Mar.–Aug.	*Saguinus* spp., *Callicebus moloch, Saimiri, Cebus* spp., parakeets
Combretum assimile	Nectar	July	7 spp. primates, marsupials, procyonids, many birds
Trees			
Ficus perforata	Fruit	Irregularly throughout the season	Nearly all monkeys, procyonids, numerous birds
Ficus erythrosticta	Fruit	Irregularly throughout the season	Nearly all monkeys, procyonids, numerous birds
Ficus killipii	Fruit	Irregularly throughout the season	Nearly all monkeys, procyonids, numerous birds
Quararibea cordata	Nectar	July–Aug.	*Saguinus* spp., *Cebus* spp., *Saimiri*, procyonids, marsupials, birds
Erythrina verna	Nectar, flowers	July–Aug.	*Ateles, Cebus apella*, many parrots, and other birds

One must bear in mind that the plant species listed here are found together on one soil type in one corner of Amazonia. Forests of quite different composition occur in other parts of the Manu Park, and of course over most of the rest of Amazonia. It can thus be expected that the identities of the critical species will vary between forests of different compositional type, but the disproportionate importance of a few key species is a strong possibility wherever there is a pronounced seasonal amplitude in resource production. Figs, because of their aseasonal fruiting, and palms, because of their long-lasting nuts, can be expected to play major roles as dry season resources throughout much of the Neotropics (Smythe 1970).

Maintenance of healthy populations of important wildlife species in mixed stands managed for production of timber and/or other products becomes at least a theoretical possibility with a knowledge of critical dry season resources. Under management the crucial species could be spared or even propagated to increase dry season carrying capacity. However, very little is known about how most plant species are recruited in natural stands, much less about the more complex problem of how to regulate the composition of multispecies communities. There is obviously a great deal more that must be learned before the tropical forest can be managed for multiple use. It is thus important that we get on with the task of acquiring the requisite knowledge, before time runs out and the possibility of instituting rational management schemes for the tropical forest is irrevocably lost.

Literature Cited

Aldrich-Blake, F.P.G. 1970. Problems of social structure in forest monkeys. In: *Social behavior in birds and mammals*. J. H. Crook, ed., pp. 79–102. Academic Press, London.

———. 1980. Long-tailed macaques. In: *Malayan forest primates: ten years' study in tropical rainforest*. D. J. Chivers, ed., pp. 147–165. Plenum Press, N.Y.

Alexander, R. D. 1974. The evolution of social behavior. *Ann. Rev. Ecol. Syst.* 5:325–383.

Altmann, S. A. 1974. Baboons, space, time and energy. *Amer. Zool.* 14:221–248.

Altmann, S. A., and J. Altmann. 1970. *Baboon ecology: African field research*. Univ. Chicago Press, Chicago.

Anderson, M. 1978. Optimal foraging: size and allocation of search effort. *Theoret. Pop. Biol.* 13:397–409.

Ashmole, N. P. 1968. Body size, prey size, and ecological segregation in five sympatric terns (Aves: Laridae). *Syst. Zool.* 17:292–304.

Baird, J. W. 1980. The selection and use of fruit by birds in an eastern forest. *Wilson Bull.* 92:63–73.

Bell, R.H.V. 1970. The use of the herb layer by grazing ungulates in the Serengeti. In: *Animal populations in relation to their food resources*. Watson, ed., pp. 111–124. Blackwell, Oxford.

Bertram, B.C.R. 1978. Living in groups: predators and prey. In: *Behavioural ecology: an evolutionary approach*. J. R. Krebs and N. B. Davies, eds., pp. 64–96. Sinauer Assoc., Inc., Sunderland, Mass.

Booth, A. H. 1956. The distribution of primates in the Gold Coast. *Journ. West. Afr. Sci. Assn.* 2:122–133.

Brown, J. L. 1964. The evolution of diversity in avian territorial systems. *Wilson Bull.* 76:160–168.

Brown, J. L., and G. H. Orians. 1970. Spacing patterns in mobile animals. *Ann. Rev. Ecol. Syst.* 1:239–262.

Brown, L., and D. Amadon. 1968. *Eagles, hawks and falcons of the world*. McGraw-Hill, N.Y.

Buskirk, R. E., and W. H. Buskirk. 1976. Changes in arthropod abundance in a highland Costa Rica forest. *Amer. Midland Natur.* 95:288–298.

Caraco, T. 1979. Time budgeting and group size: a theory. *Ecology* 60:611–617.

Carpenter, F. L., and R. E. MacMillen. 1976. Threshold model of feeding territoriality and test with a Hawaiian honeycreeper. *Science* 194:639–642.

Castro, R., and P. Soini. 1977. Field studies on *Saguinus mystax* and other callitrichids in Amazonian Peru. In: *The biology and conservation of the callitrichidae*. D. Kleiman, ed., pp. 73–78. Smithsonian Inst. Press, Washington, D.C.

Chapin, J. P. 1939. Birds of the Belgian Congo, Part II. *Bull. Amer. Mus. Nat. Hist.* 75:1–352.

Charles-Dominique, P. 1974. Ecology and feeding behavior of five sympatric Lorisids in Gabon. In: *Prosimian biology*. R. G. Martin, G. A. Doyle, and A. C. Walker, eds., pp. 135–150. Duckworth, London.

———. 1977. *Ecology and behavior of nocturnal primates: prosimians of equatorial West Africa*. Columbia Univ. Press, N.Y.

Charles-Dominique, P., and C. M. Hladik. 1971. Le *Lepilemur* du Sud de Madagascar: écologie, alimentation et vie sociale. *Terre et Vie* 25:3–66.

Charnov, E. L. 1976. Optimal foraging: the marginal value theorem. *Theor. Pop. Biol.* 9:129–136.

Chivers, D. J. 1973. An introduction to the socio-ecology of Malayan forest primates. In: *Comparative ecology and behavior of primates*. R. P. Michael and J. H. Crook, eds., pp. 101–146. Academic Press, London.

———. 1977. The feeding behavior of Siamang (*Symphalangus syndactylus*). In: *Primate ecology: studies of feeding and ranging behaviour in lemurs, monkeys and apes*. T. H. Clutton-Brock, ed., pp. 355–382. Academic Press, London.

Chivers, D. J. (ed.). 1980. *Malayan forest primates: ten years' study in tropical rain forest*. Plenum Press, N.Y.

Chivers, D. J., and J. J. Raemaekers. 1980. Long-term changes in behavior. In: *Malayan forest primates: ten years' study in tropical rain forest*. D. J. Chivers, ed., pp. 209–260. Plenum Press, N.Y.

Clutton-Brock, T. H. 1974. Primate social organization and ecology. *Nature* 250:539–542.

———. 1975. Feeding behavior of Red Colobus and Black and White Colobus in East Africa. *Folia Primatol.* 8:247–262.

———. 1977. Some aspects of intraspecific variation in feeding and ranging behavior in primates. In: *Primate ecology: studies of feeding and ranging behaviour in lemurs, monkeys and apes*. T. H. Clutton-Brock, ed., pp. 539–556. Academic Press, London.

Clutton-Brock, T. H., and P. H. Harvey. 1977a. Species differences in feeding and ranging behavior in primates. In: *Primate ecology: studies of feeding and ranging behaviour in lemurs, monkeys and apes*. T. H. Clutton-Brock, ed. Academic Press, London.

Clutton-Brock, T. H., and P. H. Harvey. 1977b. Primate ecology and social organization. *Journ. Zool. Lond.* 183:1–39.

———. 1978. Home range size, population density and phylogeny in primates. In: *Primate ecology and human origins: ecological influences on social organization.* I. S. Bernstein and E. O. Smith, eds., pp. 201–214. Garland STPM Press, N.Y.

Cody, M. L. 1971. Finch flocks in the Mojave desert. *Theoret. Pop. Biol.* 2:141–158.

Croat, T. B. 1975. Phenological behavior of habitat and habitat classes on Barro Colorado Island. *Biotropica* 7:270–277.

Crook, J. H. 1965. The adaptive significance of avian social organizations. *Symp. Zool. Soc. Lond.* 14:182–218.

———. 1972. Sexual selection, dimorphism, and social organization in the primates. In: *Sexual selection and the descent of man.* B. G. Campbell, ed. Aldine, Chicago.

Crook, J. H., and J. S. Gartlan. 1966. Evolution of primate societies. *Nature* 210:1200–1203.

Curio, E. 1976. *The ethology of predation.* Springer Verlag, N.Y.

Curtin, S. H. 1980. Dusky and banded leaf monkeys. In: *Malayan forest primates: ten years' study in tropical rain forest.* D. J. Chivers, ed., pp. 107–145. Plenum Press, N.Y.

Curtin, S. H., and D. J. Chivers. 1978. Leaf-eating primates of peninsular Malaysia: the siamang and the dusky leaf-monkey. In: *The ecology of arboreal folivores.* G. G. Montgomery, ed., pp. 441–464. Smithsonian Inst. Press, Washington, D.C.

Cushing, D. H., and F. R. Harden-Jones. 1968. Why do fish school? *Nature* 218:918–920.

Dawson, G. A. 1977. Composition and stability of social groups of the tamarin, *Saguinus oedipus geoffroyi*, in Panama. In: *The biology and conservation of the Callitrichidae.* D. G. Kleiman, ed., pp. 23–27. Smithsonian Inst. Press, Washington, D.C.

———. 1979. The use of time and space by the Panamanian tamarin, *Saguinus oedipus. Folia Primatol.* 31:253–284.

Denham, W. W. 1971. Energy relations and some basic properties of primate social organization. *American Anthropology* 73:77–95.

DeValois, R. L., and G. H. Jacobs. 1968. Primate color vision. *Science* 162:533–540.

———. 1971. Vision. In: *Behavior of nonhuman primates.* Vol. 3. A. M. Schrier and F. Stollnitz, eds., pp. 107–157. Academic Press, N.Y.

Diamond, J. M. 1973. Distributional ecology of New Guinea birds. *Science* 179:759–769.

Dittus, W.P.J. 1977. The social regulation of population density and age-sex distribution in the Toque monkey. *Behaviour* 63:281–322.

Dubost, G. 1979. The size of African forest Artiodactyls as determined by the vegetation structure. *African Journ. Ecology* 17:1–17.

Eisenberg, J. F. 1979. Habitat, economy, and society: some correlations and hypotheses for the Neotropical primates. In: *Primate ecology and human origins: ecological influences on social organization*. I. S. Bernstein and E. O. Smith, eds. Garland Press, N.Y.

Eisenberg, J. F., Muckenhirn, N. A., and R. Rudran. 1972. The relation between ecology and social structure in primates. *Science* 176:863–874.

Eisenberg, J. F., O'Connell, M. A., and P. V. August. 1979. Density, productivity and distribution of mammals in two Venezuelan habitats. In: *Vertebrate ecology in the northern Neotropics*. J. F. Eisenberg, ed., pp. 187–207. Smithsonian Inst. Press, Washington, D.C.

Eisenberg, J. F., and R. W. Thorington. 1973. A preliminary analysis of a Neotropical mammal fauna. *Biotropica* 5:150–161.

Emmons, L. H., Gautier-Hion, A., and G. Dubost. 1983. Community structure of the frugivorous-folivorous forest mammals of Gabon. *Journ. Zool. Lond.* 199:209–222.

Epple, Gisela. 1975. The behavior of marmoset monkeys (Callithricidae). In: *Primate behavior: developments in field and laboratory research*. Leonard A. Rosenblum, ed. 4:195–240. Academic Press, N.Y.

Estes, R. D. 1974. Social organization of the African Bovidae. In: *The behavior of ungulates and its relation to management*. V. Geist and F. Walther, eds. 1:166–205. I.U.C.N., Morges.

Fitzpatrick, J. 1978. Foraging behavior and adaptive radiation in the avian family Tyrannidae. Ph.D. thesis, Princeton University.

Fitzpatrick, J. W. 1980. Foraging behavior of Neotropical tyrant flycatchers. *Condor* 82:43–57.

Fleagle, J. G. 1980. Locomotion and posture. In: *Malayan forest primates: ten years' study in tropical rain forest*. D. J. Chivers, ed., pp. 191–207. Plenum Press, N.Y.

Fontaine, R. 1980. Observations on the foraging association of double-toothed kites and white-faced capuchin monkeys. *Auk* 97:94–98.

Foster, R. B. 1973. Seasonality of fruit production and seedfall in a lowland forest ecosystem in Panama. Ph.D. thesis, Duke University, Durham, N.C.

Frankie, G. W., Baker, H. G., and P. A. Opler. 1974. Comparative phenological studies of trees in tropical wet and dry forests in the lowlands of Costa Rica. *Journ. Ecol.* 62:881–919.

Freese, C. 1975. Final report: a census of non-human primates in Peru. PAHO project AMRO-0719. Unpublished report.

Gartlan, J. S., and T. T. Struhsaker. 1972. Polyspecific associations and niche separation of rain-forest anthropoids in Cameroon, West Africa. *J. Zool. Lond.* 168:221–226.

Gautier, J. P., and A. Gautier-Hion. 1969. Les associations polyspécifiques chez les Cercopithecidae du Gabon. *Terre et Vie* 2:164–201.

Gautier-Hion, A. 1970. L'organisation sociale d'une bande de Talapoins dans le N-E du Gabon. *Folia Primatol.* 12:116–141.

———. 1971. L'écologie du Talapoin du Gabon. *Terre et Vie* 25:427–490.

———. 1973. Social and ecological features of Talapoin Monkeys: comparisons with sympatric Cercopithecines. In: *Comparative ecology and behavior of primates.* M. P. Richard and J. H. Crook, eds., pp. 147–170. Academic Press, London.

———. 1978. Food niche and coexistence in sympatric primates in Gabon. In: *Recent advances in primatology.* D. J. Chivers and J. Herbert, eds., pp. 269–291. Academic Press, N.Y.

———. 1980. Seasonal variations of diet related to species and sex in a community of *Cercopithecus* monkeys. *Journ. Anim. Ecol.* 49:237–269.

Gautier-Hion, A., Emmons, L. H., and G. Dubost. 1980. A comparison of the diets of three major groups of primary consumers of Gabon (primates, squirrels and ruminants). *Oecologia* 45:182–189.

Gautier-Hion, A., and J. P. Gautier. 1974. Les associations polyspécifiques du plateau de M'passa (Gabon). *Folia Primatol.* 22:134–177.

———. 1979. Niche écologique et diversité des espèces sympatriques dans le genre *Cercopithecus. Terre et Vie* 33:493–507.

Gill, F. B., and L. L. Wolf. 1975. Economics of feeding territoriality in the golden-winged sunbird. *Ecology* 56:333–345.

Glander, K. E. 1975. Habitat description and resource utilization: a preliminary report on mantled howler monkey ecology. In: *Sociobiology and psychology of primates.* R. Tuttle, ed., pp. 36–57. World Anthropology. Mouton, The Hague.

Greenlaw, J. S. 1967. Foraging behavior of the double-toothed kite in association with white-faced monkeys. *Auk* 64:596–597.

Hall, H.R.L. 1965. Behavior and ecology of the wild patas monkey *Erythrocebus patas* in Uganda. *Journ. Zool.* 148:15–87.

Hamilton, W. D. 1971. Geometry for the selfish herd. *Journ. Theoret. Biol.* 31:295–311.

Haverschmidt, F. 1968. *Birds of Surinam.* Oliver and Boyd, Edinburgh.

Hespenheide, H. 1971. Food preference and the extent of overlap in some insectivorous birds, with special reference to Tyrannidae. *Ibis* 113:59–72.

Hladik, A., and C. M. Hladik. 1969. Rapports trophiques entre vegetation et primates dans le forêt de Barro Colorado (Panama). *Terre et Vie* 1:25–117.

Hladik, C. M. 1975. Ecology, diet, and social patterning in Old and New World primates. In: *Socioecology and psychology of primates*. R. Tuttle, ed., pp. 3–35. World Anthropology. Mouton, The Hague.

———. 1977. A comparative study of the feeding strategies of two sympatric species of leaf monkeys: *Presbytis senex* and *Presbytis entellus*. In: *Primate ecology: studies of feeding and ranging behaviour in lemurs, monkeys and apes*. T. H. Clutton-Brock, ed., pp. 323–353. Academic Press, London.

———. 1979. Diet and ecology of prosimians. In: *The study of prosimian behavior*. G. A. Doyle and R. D. Martin, eds., pp. 307–357. Academic Press, N.Y.

Hladik, C. M., and P. Charles-Dominique. 1974. The behavior and ecology of the sportive lemur (*Lepilemur mustelinus*) in relation to its dietary peculiarities. In: *Prosimian biology*. R. D. Martin, G. A. Doyle, and A. C. Walker, eds., pp. 23–47. Duckworth, London.

Hladik, C. M., and A. Hladik. 1972. Disponibilités alimentaires et domaines vitaux des primates à Ceylon. *Terre et Vie* 26:149–215.

Hladik, C. M., Hladik, A., Bousset, J., Valdebouze, P., Viroben, G., and J. Delort-Laval. 1971. Le régime alimentaire des primates de l'île de Barro Colorado (Panama). *Folia Primatol.* 16:85–122.

Holdridge, L. R. 1967. *Life zone ecology*. Tropical Science Center, San José, Costa Rica.

Holmes, R. T., and F. A. Pitelka. 1968. Food overlap among coexisting sandpipers on northern Alaskan tundra. *Syst. Zool.* 17:305–318.

Horn, H. S. 1968. The adaptive significance of colonial nesting in the Brewer's Blackbird (*Euphagus cyanocephalus*). *Ecology* 49:682–694.

Hornocker, M. G. 1969. Winter territoriality in mountain lions. *Journ. Wildl. Mgmt.* 33:457–464.

Howe, H. F. 1979. Fear and frugivory. *Amer. Natur.* 114:925–931.

Izawa, K. 1979. Foods and feeding behavior of wild Black-capped Capuchin (*Cebus apella*). *Primates* 20:57–76.

Izawa, K., and A. Mizuno. 1977. Palm fruit cracking behavior of wild Black-capped Capuchin (*Cebus apella*). *Primates* 18:773–792.

Janson, C. H. 1975. Ecology and population densities of primates in a Peruvian rainforest. Undergraduate thesis, Princeton University.

Janson, C. H., Terborgh, J., and L. H. Emmons. 1981. Non-flying mammals as pollinating agents in the Amazonian forest. *Biotropica*, suppl. 1–6.

Janzen, D. H. 1970. Herbivores and the number of tree species in tropical forests. *Amer. Natur.* 104:501–528.

Janzen, D. H. 1973a. Sweep samples of tropical foliage insects: descriptions of study sites, with data on species abundances and size distributions. *Ecology* 54:659–686.

———. 1973b. Sweep samples of tropical foliage insects: effects of seasons, vegetation types, time of day and insularity. *Ecology* 54:687–708.

———. 1979. How to be a fig. *Ann. Rev. Ecol. Syst.* 10:13–51.

Jarmon, P. 1974. The social organization of antelope in relation to their ecology. *Behavior* 48:215–267.

Jolly, A. 1972. *The evolution of primate behavior.* Macmillan Co., N.Y.

Kay, R. F., and W. L. Hylander. 1978. The dental structure of mammalian folivores with special reference to primates and *Phalangeroidea* (Marsupalia). In: *The ecology of arboreal folivores.* G. G. Montgomery, ed., pp. 173–191. Smithsonian Inst. Press, Washington, D.C.

Kiltie, R. 1980. Seed predation and group size in rain forest peccaries. Ph.D. thesis, Princeton University.

———. 1981. Application of search theory to the analysis of prey aggregation as an antipredation tactic. *Journ. Theoret. Biol.* 87:201–206.

Kiltie, R. A., and J. Terborgh. 1983. Observations of the behavior of rain forest peccaries in Peru: why do white-lipped peccaries form herds? *Zeitschr. Tierpsychol.* (in press).

Kinzey, W. G. 1974. Ceboid models for the evolution of hominoid dentition. *Journ. Human Evolution* 3:193–203.

Kinzey, W. G., and A. H. Gentry. 1979. Habitat utilization in two species of *Callicebus.* In: *Primate ecology: problem oriented field studies.* R. W. Sussman, ed., pp. 89–100. Wiley and Sons, N.Y.

Kleiman, D. G. 1977. Monogamy in mammals. *Quart. Rev. Biol.* 52:39–69.

Klein, L. L., and D. Klein. 1973. Observations on two types of Neotropical primate intertaxa associations. *Amer. Journ. Phys. Anthropology* 38:649–653.

———. 1975. Social and ecological contrasts between four taxa of Neotropical primates. In: *Sociobiology and psychology of primates.* R. Tuttle, ed., pp. 59–85. World Anthropology. Mouton, The Hague.

———. 1977. Feeding behavior of the Colombian spider monkey. In: *Primate ecology: studies of feeding and ranging behaviour in lemurs, monkeys and apes.* T. H. Clutton-Brock, ed., pp. 153–181. Academic Press, London.

Konishi, M. 1973. How the owl tracks its prey. *Amer. Sci.* 61:414–424.

Krebs, J. R. 1974. Colonial nesting and social feeding as strategies for exploiting food resources in the Great Blue Heron (*Ardea herodias*). *Behaviour* 51:99–134.

———. 1978. Optimal foraging: decision rules for predators. In: *Behavioral ecology: an evolutionary approach*. J. R. Krebs and N. B. Davies, eds., pp. 23–63. Sinauer Assoc., Inc., Sunderland, Mass.

Kummer, H. 1968. Social organization of Hamadryas baboons. *Bibliotheca Primatol*. 6:1–189.

———. 1971. *Primate societies: group techniques of ecological adaptation*. Aldine-Atherton, Chicago.

Lazarus, J. 1972. Natural selection and the functions of flocking in birds: a reply to Merton. *Ibis* 114:556–558.

Leigh, E. G., Jr., and N. Smythe. 1978. Leaf production, leaf consumption, and the regulation of folivory on Barro Colorado Island. In: *The ecology of arboreal folivores*. G. G. Montgomery, ed., pp. 51–73. Smithsonian Inst. Press, Washington, D.C.

Leighton, M., and D. R. Leighton. 1982. The relationship of size of feeding aggregate to size of food patch: howler monkeys (*Alouatta palliata*) feeding in *Trichilia cipo* fruit trees on Barro Colorado Island. *Biotropica* 14:81–90.

MacArthur, R. H., and E. Pianka. 1966. On optimal use of a patchy environment. *Amer. Natur*. 100:603–609.

MacBride, J. F. 1936. The flora of Peru. *Field Mus. Natu. Hist. Bot. Ser*. 13.

MacKinnon, J. R. 1978. Comparative feeding ecology of six sympatric primate species in West Malaysia. *Rec. Adv. Primatol*. 1:305–321.

MacKinnon, J. R., and K. S. MacKinnon. 1980. Niche differentiation in a primate community. In: *Malayan forest primates: ten years' study in tropical rain forest*. D. J. Chivers, ed., pp. 167–190. Plenum Press, N.Y.

McMahon, T. 1973. Size and shape in biology. *Science* 179:1201–1204.

McNab, B. K. 1963. Bioenergetics and the determination of home range size. *Amer. Natur*. 97:133-140.

———. 1978. Energetics of arboreal folivores: physiological problems and ecological consequences of feeding on an ubiquitous food supply. In: *The ecology of arboreal folivores*. G. G. Montgomery, ed., pp. 153–162. Smithsonian Inst. Press, Washington, D.C.

———. 1980. Food habits, energetics, and the population biology of mammals. *Amer. Natur*. 116:106–124.

Mason, W. A. 1968. Use of space by *Callicebus* groups. In: *Primates: studies in adaptation and variability*. P. Jay, ed., pp. 200–216. Holt, Rinehart and Winston, N.Y.

Milton, K. 1978. Behavioral adaptations to leaf-eating in the mantled howler monkey. In: *The ecology of arboreal folivores*. G. G. Montgomery, ed., pp. 535–549. Smithsonian Inst. Press, Washington, D.C.

———. 1979. Factors influencing leaf choice by howler monkeys: a test of some hypotheses of food selection by generalist herbivores. *Amer. Natur.* 114:362–377.

———. 1980. *The foraging strategy of howler monkeys: a study in primate economics.* Columbia Univ. Press, N.Y.

———. 1981. Food choice and digestive strategies of two sympatric primate species. *Amer. Natur.* 117:496–505.

Milton, K., and M. L. May. 1976. Body weight, diet and home range area in primates. *Nature* 259:459–462.

Mitani, J. C., and P. S. Rodman. 1979. Territoriality: the relation of ranging pattern and home range size to defendability, with an analysis of territoriality among primate species. *Behav. Ecol. Sociobiol.* 5:241–251.

Mittermeier, R. A., and M.G.M. van Roosmalen. 1981. Preliminary observations on habitat utilization and diet in eight Surinam monkeys. *Folia Primatol.* 36:1–39.

Montgomery, G. G., and M. E. Sunquist. 1975. Impact of sloths on Neotropical forest energy flow and nutrient cycling. In: *Tropical ecological systems: trends in terrestrial and aquatic research.* F. B. Golley and E. Medina, eds., pp. 69–98. Springer Verlag, N.Y.

Morrison, D. W. 1978a. Foraging ecology and energetics of the frugivorous bat *Artibeus jamaicensis. Ecology* 59:716–723.

———. 1978b. Lunar phobia in a Neotropical fruit bat, *Artibeus jamaicensis. Anim. Behav.* 26:852–855.

Morse, D. H. 1970. Ecological aspects of some mixed-species foraging flocks of birds. *Ecolog. Monogr.* 40:119–168.

———. 1977. Feeding behavior and predator avoidance in heterospecific groups. *BioScience* 27:332–339.

Munn, C. A., and J. W. Terborgh. 1980. Multi-species territoriality in Neotropical foraging flocks. *Condor* 81:338–347.

Nagy, K. A., and K. Milton. 1979. Energy metabolism and food consumption by wild howler monkeys (*Alouatta palliata*). *Ecology* 60:475–480.

Neill, S. R., and J. M. Cullen. 1974. Experiments on whether schooling by their prey affects the hunting behavior of cephalopods and fish predators. *Journ. Zool. Lond.* 172:549–569.

Neyman, P. F. 1977. Some aspects of the biology of free-ranging cotton-top tamarins (*Saguinus o. oedipus*) and conservation status of the species. In: *The biology and conservation of the Callitrichidae.* D. G. Kleiman, ed., pp. 39–71. Smithsonian Inst. Press, Washington, D.C.

Oates, J. F. 1977. The guereza and its food. In: *Primate ecology: studies of feeding and ranging behaviour in lemurs, monkeys and apes.* T. H. Clutton-Brock, ed., pp. 275–321. Academic Press, London.

Oppenheimer, J. R. 1977. *Presbytis entellus*, the Hanuman Langur. In: *Pri-*

mate conservation. Rainier III of Monaco and G. H. Bourne, eds., pp. 469–512. Academic Press, N.Y.

Orians, G. H. 1969a. On the evolution of mating systems in birds and mammals. *Amer. Natur.* 103:589–603.

———. 1969b. The number of bird species in some tropical forests. *Ecology* 50:783–801.

Orians, G. H., and N. E. Pearson. 1979. On the theory of central place foraging. In: *Analysis of ecological systems*. D. J. Horn, R. D. Mitchell, and G. R. Stairs, eds., pp. 155–177. Ohio State Univ. Press, Columbus.

Papageorgis, C. 1975. Mimicry in Neotropical butterflies. *Amer. Sci.* 63:522–532.

Payne, R. S. 1971. Acoustic location of prey by barn owls (*Tyoto alba*). *Journ. Exp. Biol.* 54:535–573.

Pearson, D. L. 1971. Vertical stratification of birds in a tropical dry forest. *Condor* 73:46–55.

———. 1977. A pantropical comparison of bird community structure on six lowland forest sites. *Condor* 79:232–244.

Pollock, J. I. 1979. Spatial distribution and ranging behavior in lemurs. In: *The study of prosimian behavior*. G. A. Doyle and R. D. Martin, eds., pp. 359–409. Academic Press, N.Y.

Pook, A. G., and G. Pook. 1981. A field study of the socio-ecology of the Goeldi's Monkey (*Callimico goeldii*) in northern Bolivia. *Folia Primatol.* 35:288–312.

Pough, F. H. 1973. Lizard energetics and diet. *Ecology* 54:837–844.

Powell, G.V.N. 1974. Experimental analysis of the social value of flocking by starlings (*Sturnis vulgaris*) in relation to predation and foraging. *Anim. Behav.* 22:501–505.

Pulliam, H. R. 1973. On the advantage of flocking. *Journ. Theoret. Biol.* 38:419–422.

———. 1976. The principle of optimal behavior and the theory of communities. In: *Perspectives in ethology*. P.P.G. Bateson and P. H. Klopfer, eds. 2:311–332. Plenum Press, N.Y.

Pulliam, H. R., Anderson, K. A., Misztal, A., and N. Moore. 1974. Temperature-dependent social behavior in juncos. *Ibis* 116:360–364.

Pulliam, H. R., Caraco, T., and S. Martindale. 1980. Avian flocking in the presence of a predator. *Nature* 285:400–401.

Pyke, G. H., Pulliam, H. R., and E. L. Charnov. 1977. Optimal foraging: a selective review of theory and tests. *Quart. Rev. Biol.* 52:137–153.

Raemaekers, J. J., Aldrich-Blake, F.P.G., and J. B. Payne. 1980. The forest. In: *Malayan forest primates: ten years' study in tropical rain forest*. D. J. Chivers, ed., pp. 29–61.

Raemaekers, J. J., and D. J. Chivers. 1980. Socio-ecology of Malayan forest primates. In: *Malayan forest primates: ten years' study in tropical rain forest*. D. J. Chivers, ed., pp. 279–316. Plenum Press, N.Y.

Ramirez, M. 1980. Grouping patterns of the woolly monkey, *Lagothrix lagotricha*, at the Manu National Park, Peru. *Amer. Journ. Phys. Anthro.* 52:269.

Ramirez, M. F., Freese, C. H., and J. Revilla C. 1977. Feeding ecology of the Pygmy Marmoset, *Cebuella pygmaea*, in northeastern Peru. In: *The biology and conservation of the Callitrichidae*. D. Kleiman, ed., pp. 91–104. Smithsonian Inst. Press, Washington, D.C.

Rettig, N. L. 1978. Breeding behavior of the Harpy Eagle (*Harpia harpyja*). *Auk* 95:629–643.

Richards, P. W. 1952. *The tropical rainforest*. Cambridge Univ. Press, Cambridge.

Ricklefs, R. E. 1975. Seasonal occurence of night flying insects on Barro Colorado Island. *Journ. New York Entomol. Soc.* 83:19–32.

Robertson, D. R., Sweatman, H. P., Fletcher, E. A., and M. G. Cleland. 1976. Schooling as a mechanism for circumventing the territoriality of competitors. *Ecology* 57:1208–1220.

Robinson, M. H., and B. Robinson. 1970. Prey caught by a sample population of the spider *Argiope argentata* (Araneae: Araneidae) in Panama: a year's census data. *Zool. Journ. Linn. Soc.* 49:345–358.

Rodman, P. S. 1977. Feeding behavior of Orang-utans of the Kutai Nature Reserve, East Kalimantan. In: *Primate ecology: studies of feeding and ranging behaviour in lemurs, monkeys and apes*. T. H. Clutton-Brock, ed., pp. 383–413. Academic Press, London.

———. 1978. Diets, densities, and distributions of Bornean primates. In: *The ecology of arboreal folivores*. G. G. Montgomery, ed., pp. 465–478. Smithsonian Inst. Press, Washington, D.C.

Rudran, R. 1978. Socioecology of the blue monkeys (*Cercopithecus mitis stuhlmanni*) of the Kibale Forest, Uganda. *Smithsonian Contrib. Zool.* 249:1–88.

———. 1979. The demography and social mobility of a Red Howler (*Alouatta seniculus*) population in Venezuela. In: *Vertebrate ecology in the northern Neotropics*. J. F. Eisenberg, ed., pp. 107–126. Smithsonian Inst. Press, Washington, D.C.

Schaller, G. B. 1972. *The Serengeti lion: a study of predator-prey relations*. Univ. of Chicago Press, Chicago.

Schaller, G. B., and P. G. Crawshaw, Jr. 1980. Movement patterns of jaguar. *Biotropica* 12:161–168.

Schoener, T. W. 1967. The ecological significance of sexual dimorphism in size in the lizard *Anolis conspersus*. *Science* 155:474–477.

———. 1968. The *Anolis* lizards of Bimini: resource partitioning in a complex fauna. *Ecology* 49:704–726.

———. 1971. Theory of feeding strategies. *Ann. Rev. Ecol. Syst.* 2:369–404.

Schoener, T. W., and D. H. Janzen. 1968. Notes on environmental determinants of tropical versus temperate insect size patterns. *Amer. Natur.* 102:207–224.

Sherman, P. W. 1977. Nepotism and the evolution of alarm calls. *Science* 197:1246–1253.

Siegfried, W. R., and L. G. Underhill. 1975. Flocking as an antipredator strategy in doves. *Anim. Behav.* 23:504–508.

Silberbauer-Gottsberger, I., Morawetz, W., and G. Gottsberger. 1977. Frost damage of cerrado plants in Botucatu, Brazil, as related to the geographical distribution of the species. *Biotropica* 9:253–261.

Smith, C. C. 1977. Feeding behavior and social organization in howling monkeys. In: *Primate ecology: studies of feeding and ranging behaviour in lemurs, monkeys and apes.* T. H. Clutton-Brock, ed., pp. 97–126. Academic Press, London.

Smythe, N. 1970. Relationships between fruiting seasons and seed dispersal methods in a Neotropical forest. *Amer. Natur.* 104:25–35.

———. 1974. Biological monitoring data: insects. In: *1973 environmental monitoring and baseline data.* R. W. Rubinoff, ed., pp. 70–115. Smithsonian Inst. Environmental Sci. Prog., Washington, D.C.

Snodderly, D. M., Jr. 1972. In: *Perspectives in Primate Biology.* A. B. Chiarelli, ed., pp. 93–149. Plenum Press, N.Y.

———. 1979. Visual discriminations encountered in food foraging by a Neotropical primate: Implications for the evolution of color vision. In: *The behavioral significance of color.* E. H. Burtt, Jr., ed., pp. 238–279. Garland STPM Press, N.Y.

Stiles, E. 1980. Patterns of fruit presentation and seed dispersal in bird-disseminated woody plants in the eastern deciduous forest. *Amer. Natur.* 116:670–688.

Storer, R. W. 1966. Sexual dimorphism and food habits in three North American accipiters. *Auk* 83:423–435.

Stott, K. 1947. Fairy Bluebird-Long-tailed Macaque association on Mindanao. *Auk* 64:130.

Struhsaker, T. T., and L. Leland. 1977. Palm-nut smashing by *Cebus a. apella* in Colombia. *Biotropica* 9:124–126.

———. 1979. Socio-ecology of five sympatric monkey species in the Kibale Forest, Uganda. *Adv. Stud. Behav.* 9:159–227.

Sunquist, M. E. 1981. The social organization of tigers (*Panthera tigris*) in

Royal Chitawan National Park, Nepal. *Smithsonian Contrib. Zool.* 336:1–98.

Sussman, R. W. 1974. Ecological distinctions in sympatric species of lemur. In: *Prosimian biology*. R. Martin, G. A. Doyle, and A. C. Walker, eds., pp. 75–108. Duckworth, London.

———. 1977. Feeding behavior of *Lemur catta* and *Lemur fulvus*. In: *Primate ecology: studies of feeding and ranging behaviour in lemurs, monkeys and apes*. T. H. Clutton-Brock, ed., pp. 1–36. Academic Press, London.

———. 1979. Nectar-feeding by prosimians and its evolutionary and ecological implications. *Recent Adv. Primatol.* 2:569–577.

Terborgh, J. 1977. Bird species diversity on an Andean elevational gradient. *Ecology* 58:1007–1019.

———. 1980. Causes of tropical species diversity. *Actis XVII Congr. Internat. Ornithol.*, pp. 955–961.

Terborgh, J., and J. M. Diamond. 1970. Niche overlap in feeding assemblages of New Guinea birds. *Wilson Bull.* 82:29–52.

Terwilliger, V. J. 1978. Natural history of Baird's Tapir on Barro Colorado Island, Panama Canal Zone. *Biotropica* 10:211–220.

Thorington, R. W., Jr. 1967. Feeding and activity of *Cebus* and *Saimiri* in a Colombian forest. In: *Neue Ergebnisse der Primatologie*. D. Stark, R. Schneider, and H. J. Kuhn, eds., pp. 180–184. Fischer Verlag, Stuttgart.

Vine, I. 1971. Risk of visual detection and pursuits by a predator and the selective advantage of flocking behavior. *J. Theoret. Biol.* 30:405–422.

Waser, P. M. 1975a. Monthly variations in feeding and activity patterns of the mangabey, *Cercocebus albigena* (Lyddeker). *East Afr. Wildlife Journ.* 13:249–263.

———. 1975b. Experimental playbacks show vocal mediation of avoidance in a forest monkey. *Nature* 255:56–58.

———. 1976. *Cercocebus albigena*: site attachment, avoidance, and intergroup spacing. *Amer. Natur.* 110:911–935.

———. 1977. Feeding, ranging and group size in the mangabey *Cercocebus albigena*. In: *Primate ecology: studies of feeding and ranging behaviour in lemurs, monkeys and apes*. T. H. Clutton-Brock, ed. Academic Press, London.

Waser, P. M., and T. J. Case. 1981. Monkeys and matrices: coexistence of "omnivorous" forest primates. *Oecologia* 49:102–108.

Werner, E. E. 1977. Species packing and niche complementarity in three sunfishes. *Amer. Natur.* 111:553–578.

Wiley, R. H. 1971. Cooperative roles in mixed flocks of antwrens (*Formicariidae*). *Auk* 88:881–892.

Willis, E. O., and Y. Oniki. 1978. Birds and army ants. *Ann. Rev. Ecol. Syst.* 9:243–264.

Wilson, E. O. 1975. *Sociobiology: the new synthesis.* Harvard Univ. Press, Cambridge, Mass.

Windsor, D. M. 1976. *Environmental monitoring and baseline data.* Smithsonian Inst. Press, Washington, D.C.

Wittenberger, J. F. 1980. Group size and polygamy in social mammals. *Amer. Natur.* 115:197–222.

Wolda, H. 1978. Seasonal fluctuations in rainfall, food and abundance of tropical insects. *Journ. Anim. Ecol.* 47:369–381.

Wolda, H., and F. W. Fisk. 1981. Seasonality of tropical insects. II. Blattaria in Panama. *Journ. Anim. Ecol.* 50:827–838.

Woolfenden, G. E., and J. W. Fitzpatrick. 1978. The inheritance of territory in group-breeding birds. *BioScience* 28:104–108.

Wrangham, R. W. 1977. Feeding behavior of chimpanzees in Gombe National Park, Tanzania. In: *Primate ecology: studies of feeding and ranging behaviour in lemurs, monkeys and apes.* T. H. Clutton-Brock, ed., pp. 503–538. Academic Press, London.

———. 1980. An ecological model of female-bonded primate groups. *Behaviour* 75:262–300.

Wright, P. C. 1978. Home range, activity pattern, and agonistic encounters of a group of night monkeys (*Aotus trivirgatus*) in Peru. *Folia Primatol.* 29:43–55.

Yoneda, M. 1981. Ecological studies of *Saguinus fuscicollis* and *Saguinus labiatus* with reference to habitat segregation and height preference. *Kyoto University Overseas Research Reports of New World Monkeys* 2:43–50.

Author Index

Subject Index

Library of Congress Cataloging in Publication Data

Terborgh, John, 1936-
Five New World primates.
(Monographs in behavior and ecology)
Bibliography: p. Includes index.
1. Cebidae—Ecology. 2. Mammals—Ecology.
3. Mammals—Peru—Ecology. I. Title.
II. Title: 5 New World primates. III. Series.
QL737.P925T46 1983 599.8'2 83-42596
ISBN 0-691-08337-1
ISBN 0-691-08338-X (pbk.)

John Terborgh is Professor of Biology at Princeton University.